T0138938

Tapping Molecular Wilderness

Tapping Molecular Wilderness

Drugs from Chemistry–Biology–Biodiversity
Interface

Yongyuth Yuthavong

PAN STANFORD PUBLISHING

Published by

Pan Stanford Publishing Pte. Ltd.
Penthouse Level, Suntec Tower 3
8 Temasek Boulevard
Singapore 038988

Email: editorial@panstanford.com
Web: www.panstanford.com

British Library Cataloguing-in-Publication Data
A catalogue record for this book is available from the British Library.

ISBN 978-981-4613-59-0 (Hardcover)
ISBN 978-981-4316-60-6 (eBook)

Printed in the USA

Reviews

"Writing a popular science book is more challenging than writing a professional one for the technical audience. One needs to be scientifically rigorous, yet speak in the language of the school student and the 'lay' public. There can be no threatening equations or complex chemical pathways, yet one should convey the message in a lucid manner. Professor Yuthavong carries it off with ease and elan. He has chosen the word "wilderness" deliberately, to evoke both excitement and awe in the reader. He shows how human creativity is able to chisel molecules from wilderness into useful products, how nature itself has been doing such molecular architecture over evolution, and how we may learn from it. The underlying message, expressed with time honored wisdom, is Gandhian in spirit. Recall what Mahatma Gandhi said: 'Nature provides for man's need, but not his greed', and 'Be the change you want the world to be'. This is a book that needs to be distributed across both the developing and developed worlds."

D. Balasubramanian
Professor and Director of Research,
L V Prasad Eye Institute, Hyderabad, India, and
UNESCO Kalinga Prize Laureate in Science Popularization

"I very much like the idea of writing something that's technically correct but intended for a general audience. The topics would correct an impression that all drug discovery these days comes from high throughput screening of synthetic molecules. I'm very impressed with the variety of topics the writer has managed to touch upon and with how technically accurate the handling of these topics has been."

Jon Clardy
Professor, Harvard Medical School and Broad Institute, USA

"This pioneering book is a powerful source of enlightenment on the vital connections between the diversity world's biological splendour

and advancement of scientific knowledge. It offers a convincing case as to why the conservation of biological diversity is imperative for human wellbeing. I recommend it to anyone who has an interest in sustainable development in general and environmental protection in particular."

Calestous Juma
Professor, Harvard Kennedy School, USA, and
Former Executive Secretary, United Nations Convention on
Biological Diversity

"This is an excellent reading not only for researchers and students but also for general readers. The whole book is woven around the key term 'wilderness'. It covers a wide area of subjects, from ancient myth to modern molecular biology and drug design. The book is not only educational but also highly entertaining. I hope in the future it will be available to those people who do not understand English."

Hisao Masai
Professor, Tokyo Metropolitan Institute of Science
and University of Tokyo, Japan

"The need to bring together new knowledge in basic sciences, agriculture, anecdotes and cultural norms on a single platform for efforts in prospecting for drugs from natural products cannot be overemphasized. Many have attempted to do this but only a few have the background necessary to succeed in the efforts. Professor Yongyuth brings with him a wealth of knowledge accumulated over thirty years and is probably the best to produce a much needed balanced view in the field."

Ayoade Oduola
Former Deputy Director,
UNDP/World Bank/WHO Special Programme for
Research and Training in Tropical Diseases, Geneva, Switzerland

"Professor Yongyuth Yuthavong has worked for decades at the highest levels of science and government and successfully cross pollinated these worlds. So it's no surprise that his new book, Tapping Molecular Wilderness: Drugs from Chemistry–Biology–Biodiversity Interface,

bridges the worlds of science and nature. Coming at the moment when the world is embarking on a new set of Sustainable Development Goals which also must embrace both science and nature, Prof. Yuthavong's book can be widely recommended for anyone who wishes to think more deeply about these goals—and the future of our world."

Peter Singer
Professor, University of Toronto, and
Chief Executive Officer, Grand Challenges Canada

"One thing that typifies the writer is his clarity in thinking and presentation: This quality is apparent in this highly readable book. Through hands-on drug research and involvement with related issues, he aims to make us appreciate nature for its cornucopia of simple and complex molecules that are beneficial to mankind. One such benefit is the natural products for combating pathogenic organisms whose drug resistance should be taken seriously by our making sustained and renewed efforts to fight them. After all pathogens must fight for their lives; simplistic and ephemeral efforts by the medical community have constantly proved to be inadequate. In this book the themes of the need to sustain nature for its biodiversity and to combat pathogens by natural and modified biomolecules shine through brilliantly."

Bhinyo Panijpan
Former Director, Institute for Innovative Learning,
Mahidol University, Thailand

"The author beautifully portraits the biodiverse 'molecular wilderness' as the world of wonder, full of treasure to benefit mankind. Complex chemistry of drug discovery and drug design is amazingly made simple. It ends with a strong message that molecular wilderness is powerful. We must respect its balance and coexist with it sustainably. Otherwise it fights back harshly. The book is very educational and inspiring. It is a complex scientific textbook neatly made simple for general readers. We definitely need more science and technology books in this literary style."

Khunying Sumonta Promboon
Member of Thai National Legislative Assembly and
Former President of Srinakarinwirote University, Thailand

"Living organisms produce both toxic compounds to disable their predators and beneficial compounds to protect or heal themselves, so as to enhance their ability to survive. So Nature, or the 'Wilderness', is a rich source of medically important molecules. Thus 'Tapping Molecular Wilderness' has played a crucial role in the discovery of new drugs to combat human illnesses, such as infection and heart disease. The author elegantly discusses the principles of drug discovery, the need for an integrated role of chemistry and biology, novel strategies in research, as well as problems arising from drug resistance. As expert researcher, with success in devising a novel drug for malaria, the author has simplified the scientific concepts, historical perspectives and modern trends in drug discovery in a simplified manner, readily understood by the layman. More books like this are needed to show the importance of research, not only at applied level but also at basic level: Perhaps then governments, especially in developing countries, may invest more in research for the future."

M. R. Jisnuson Svasti
Emeritus Professor, Mahidol University and Chulaborn Research
Institute, Thailand

"The author should be admired for his bold effort to write a book on 'natural science' for the general public. As it turns out, this book not only contains a wealth of scientific information but also is very easy to read and to follow from the first page to the last. Readers will benefit from the knowledge given which can be used as a starting point to dig further into the 'beauty of nature'. The author should be congratulated for the beautiful tale of science adventure."

Yodhathai Thebtaranonth
Emeritus Professor, Mahidol University, Thailand, and
ASEAN Outstanding Technologist and Technologist Awardee, 1995

"From the wilderness have come many revelations. Professor Yongyuth Yuthavong now has added chemistry to the list."

Prapon Wilairat
Professor, Mahidol University, Thailand, and
Outstanding Scientist of Thailand Awardee, 1997

Contents

Preface

This is a book for general readers with some background in science, concerning the search for drugs, starting from molecular diversity found in nature, which might be called molecular wilderness. The drug molecules may be used as such, or may be used as templates for synthetic or semi-synthetic drugs obtained from the interface of chemistry, biology and biodiversity. In some cases, the active parts from natural molecules may be identified and modified to more effective ones. In other cases, nature provides the targets, such as essential enzymes from infectious microorganisms, from which synthetic drugs can be designed. The mechanisms of action of drugs can be discerned from studying the target-drug interactions. Nature may fight back, as when microorganisms become resistant to drugs, but we can again use the chemistry–biology–biodiversity interface to obtain drugs which overcome the resistance. The battle goes on, hopefully with victory on the human side, but this requires special efforts from wider areas than medical science.

This book offers a bird's eye view on the unifying theme of interface between chemistry and biology as the essence of drug discovery, with focus on "conversation" between science and nature. Examples are taken from successes in discovering useful drugs from the wilderness of biodiversity, from aspirin and quinine to antibiotics and statins. Failures following initial successes due to dynamic nature of molecular wilderness are also highlighted, with examples of eventual successes. The book concentrates on early stage discovery, which requires interdisciplinary approaches combining synthetic with structural chemistry, biochemistry, molecular and cell biology, but also highlights the importance of pharmacology, toxicology, pre-clinical and clinical sciences to complete the chain of drug development. Significantly, the book draws attention on biodiversity as a key to sustainable efforts to discover new drugs from nature.

Most books on natural products and drug development concern mainly or only technical and scientific aspects of the topic. Others on environmental and indigenous knowledge tend to ignore or, worse still, tend to be hostile to the scientific approaches. This book attempts to bridge the gap between the "two cultures", hopefully resulting in balanced understanding of various issues in development of drugs from nature. It also substantially covers a hitherto little explored topic of plasticity of drug targets and the various ways in which nature "fights back" against our attempts to conquer infectious and other diseases, resulting in drug resistance, or in some cases emergence of other diseases. The conclusion is that brute technological forces alone are insufficient to solve our present problems or prevent new ones, but that science and technology have to be integrated into other aspects of health care, including social science and integrated economic and social development.

The book recalls that biodiversity contains a large number of products, many of which have been used in the form of traditional medicine, and others have been identified as drugs or drug leads for modern medicine. Yet others provide the targets for design of drugs, both based on natural sources and synthetic chemistry. It points out that poor and vulnerable populations still rely substantially on traditional medicine for their health care, the quality of which can be improved by modern science. Conversely, extension of traditional medicine through research can contribute to progress of modern medicine, leading to cheaper and more accessible drugs. The message of the book is that tapping molecular wilderness should be done responsibly, ensuring that fair benefits go back to the indigenous population where the traditional knowledge originated. It also needs to be done in an environmentally sustainable fashion through the help of science as major tools.

The wilderness has been around much longer than we have. Simple cells appeared about 3.6 billion years ago, about a billion years after the formation of our planet. Multicellular life started about a billion years ago. Land plants and animals started to appear from about half that time. In contrast, modern humans only evolved some 100,000 years ago. In this very short history of human beings, we have managed to exert enormous influence on the wilderness, taming many species for agriculture, and condemning many more to extinction by our disregard or ignorance. Until recently, we have tapped the wilderness for our own use as though it is an unlimited reservoir.

Only recently have we come to realize that the wilderness has limits to human insults, with grim consequences for our own existence. By destroying forests for their products and turning the land to our own use, we have unwittingly created deserts, drained away water resources and contributed to global warming. We have to stop the reckless behaviour, not just to be kind to the wilderness but indeed for the sake of our own survival.

The book deals with the topic of tapping the wilderness for human purposes with three distinct characters. First, it takes chemistry of nature as the essence of wilderness. It considers natural molecules as members of interacting components underlying the phenomena of wilderness. Secondly, it concentrates on tapping this molecular wilderness for drugs, both from the natural molecules themselves and from the use of these molecules as design models for synthetic drugs. Thirdly, it treats the threats of drug resistance of microbes as natural outcomes of interactions of molecular wilderness. Like natural disasters of desertification and flooding, resistance of microbes to drugs is viewed as consequences of disturbance of wilderness.

Just as we need to tap wilderness in the visible world sustainably, so do we need to tap molecular wilderness in a sustainable manner. Both are huge challenges, requiring change of mindset as well as technical progress. The world community is embarking on cooperation, on an unprecedented scale, through United Nations and other world bodies to try to achieve "Sustainable Development Goals". These will cover major issues of development in economic, social and environmental fields. The goals for sustainable tapping of molecular wilderness are different, achieving new effective drugs and overcoming the problems of drug resistance. The goals are more modest, perhaps, but no less worthwhile.

Acknowledgements

I would like to thank Bongkoch Tarnchompoo, Kritsachai Somsaman, Penchit Chitnumsub, Philip James Shaw, Sumalee Kamchonwongpaisan and other members of the National Centre for Genetic Engineering and Biotechnology (BIOTEC), Thailand National Science and Technology Development Agency (NSTDA), for their support in the writing of this book. Thanks are due to Yodhathai Thebtaranonth, Tirayut Vilaivan, Prapon Wilairat, Jisnuson Svasti, Bhinyo Panijpan, Thanat Chookajorn, Sumonta Promboon, Ayo Oduola, Ken-ichi Arai, Hisao Masai, Jon Clardy, Dyann Wirth, Frank Petersen and Dominique Charron for their comments and help in various stages of the book.

A Brief Description of the Book

The book is divided into six chapters:

1. *Molecular Wilderness, Harsh and Healing.* The biodiverse environment contains molecules both noxious and healing for humans. Natural products are chemical expressions of the molecular wilderness. Large biomolecules in living organisms are targets or receptors for smaller molecules, including man-made drugs.

2. *Gifts from Molecular Wilderness.* Humans have over the ages discovered useful remedies from herbs and other natural substances, the nature of most of which is learnt of only much later. Chemistry interfaces with biology to refine and produce drug substances from nature, often with the help of knowledge from traditional medicine, which is still the major source of provision of health care for most people in developing countries. Tapping the molecular wilderness needs to be done sustainably and responsibly, with fair benefit to the indigenous people whom we must respect for their heritage and traditional wisdom.

3. *Drug Targets from Molecular Wilderness.* Infectious diseases represent the dark side of wilderness which humankind still has to contend with, especially the diseases which affect the majority of the world's population still living in poverty. The disease pathogens have essential components or processes which can be specific targets for drugs and vaccines. A variety of approaches can be used to identify such targets, including gene knockouts and chemical genomics. Random screening of compounds already available in large pharmaceutical collections can also provide new leads, even when the targets have not yet been identified.

4. *Molecular Wilderness as Templates for Drugs.* Original molecules from nature provide templates which can be scaled up or modified to make better drugs through chemistry, biology and allied sciences. Molecular diversity from combinatorial

and diversity-oriented chemical synthesis provides even wider selections. Fragment-based drug discovery shows the power of making effective drugs from components, each of which may bind only weakly to the target. Combinatorial biosynthesis provides a method for producing "nonnatural" natural products, while metagenomics can lead to discovery of new antibiotics, even from microorganisms which cannot be grown. Rules governing the ability of drug molecules to act effectively, including surviving long enough in the host and accessing tissues and targets of action, can be used to build better platforms for development of new drugs from original natural molecules.

5. *The Wilderness Fights Back.* Microorganisms and diseased cells can develop resistance to drugs through various mechanisms. Drug resistance can be viewed as the natural tendency for the wilderness to fight back against human intervention, just as life evolves from struggle for existence in the natural world. Poor human behavior and public health practice contribute to the emergence of drug resistance. We need to understand the mechanisms of drug resistance and find rational approaches to overcome the resistance, either by modifying old drugs or by finding new drugs, including drug combinations. Good examples of natural combinations can be found from the strategies which microorganisms use to prey on others or defend themselves in the ecosystem. In addition to the spectre of drug resistance, new molecular wilderness threats are looming in the shape of emerging diseases resulting from global climate and social changes. These threats require vigilance and quick responses coordinated on the global scale.

6. *Living with Molecular Wilderness.* Learning from past lessons, we should come to realize the power of molecular wilderness, both to yield benefits for the human species, and to strike back when we oversimplify its exploitation and disregard the delicate balance of nature. Coexistence and conservation should be preferred over exploitation and subjugation of the molecular wilderness. Sustainable tapping of the molecular wilderness requires not only science and technology but a balanced approach, taking into account the social, economic and environmental factors affecting the health of people all

over the world. Furthermore, since the majority of Nature's biodiversity is contained in tropical countries, where the standards of health care are still poor, sustainable tapping of the molecular wilderness should also be done with the objective of improving these standards so as to achieve a healthy world for all.

Chapter 1

Molecular Wilderness, Harsh and Healing

... in Wildness is the preservation of the world ...

—Henry David Thoreau, *Walking*, 1862

Summary

The biodiverse environment contains molecules both noxious and healing for humans. Natural products are chemical expressions of the molecular wilderness. Large biomolecules in living organisms are targets or receptors for smaller molecules, including man-made drugs.

1.1　Wilderness Is Harsh

The wilderness spells danger. Human dwellings, be they villages or cities, are built away from, or to replace, the wilderness so that dangers lurking there can be avoided. A wild area left in its natural condition also carries inconveniences for our normal daily lives. Only on special occasions—vacations or work-related field trips, perhaps—do most people venture into untamed forests or unknown grounds.

Tapping Molecular Wilderness: Drugs from Chemistry–Biology–Biodiversity Interface
Yongyuth Yuthavong
Copyright © 2016 Pan Stanford Publishing Pte. Ltd.
ISBN 978-981-4613-59-0 (Hardcover), 978-981-4316-60-6 (eBook)
www.panstanford.com

The wilderness can be harsh. Ruggedness of the terrain is compounded by threats posed by its inhabitants. In a tropical forest, for example, we can be bitten by a snake, stung by an insect or poisoned by a seemingly edible mushroom. Away from the comfort and artificial surroundings of the modern human society, we are thrown back to the days when we were still living as savages, albeit with the help of modern tools which we bring with us. Back in the wilderness, we are surrounded by a habitat in which, as the poet Alfred Lord Tennyson [1] puts, 'nature, red in tooth and claw' prevails. The richness of biodiversity is the result of complex

---

Box 1.1 The Farmer and the Cobra

In an Aesop tale, a farmer took a cold cobra up out of pity and warmed it with his hands. Gaining movement, the cobra bit the farmer and killed him. Apart from the moral of the story, the sad truth is that a number of farmers and other people still die from poisonous snake bites. Cobras (see Fig. 1.1) and some 400 other poisonous snakes account for about 20,000 deaths from their bites every year, mostly in the tropics. Snake venoms contain many components, chiefly neurotoxins, which attack the nervous system; haemotoxins, which destroy blood cells and other tissues and disrupt blood clotting; and other components which disrupt blood pressure regulation or cause other toxic effects [12, 13]. Some of these are proteins which block neural transmission by mimicking the shape of a natural nerve-signalling molecule acetylcholine. Others are enzymes, proteins with catalytic functions, which digest cell membrane components, causing cell rupture. People who are bitten can be saved by antivenoms, which are neutralizing antibodies derived from injecting nonlethal doses of venom to a host animal such as a horse or a sheep.

The biological functions of snake venom are to immobilize and digest prey and to defend against enemies, which some cobras can do, not only by biting, but also by spitting venom to some distance.

Because of their biological effects, some snake venom components have been used as diagnostic tools or drugs. For example, a venom component from the Malayan pit viper which causes prolonged bleeding by dissolving blood clots has been found to be useful for treatment of heart disease and stroke. In another example, a protein similar to a venom protein of the Brazilian pit viper which affects blood pressure regulation has been used as a drug to lower blood pressure.

(Continued)

Box 1.1 (*Continued*)

(a)   (b)   (c)

Figure 1.1   Hazards of the wilderness. (a) Cobra (*Naja naja*), (b) the blister beetle (*Epicorta hirticornis*) and (c) a poison mushroom (*Amanita longistriata*). Pictures (a) and (b) courtesy of the National Science Museum of Thailand; picture (c) courtesy of Thitiya Boonpratuang, BIOTEC Thailand.

relationships among various organisms, where each species—prey and predator alike—thrives according to its niche. Obvious evidence of these relationships includes food chains and food webs, where animals, plants and microorganisms are eaters or eaten. Simply put, the claw and tooth of the predator are red from the blood of its prey. Leading up to this act, the predator may lure its prey with bait under the guise of food or other attractions, sting it with venom or use various means to trap and kill it. On its part, the prey can be elusive, through camouflage or other means of fooling the predator, such as mimicking something the predator hates or fears, or simply through a quick escape. When cornered by the predator, the prey fights it off with defensive weapons. The bee uses its venomous sting to defend against larger predators. The pufferfish harbours bacteria which produce highly poisonous toxin to help it to ward off its predators, thereby gaining the common benefit of survival. Many South American frogs have brilliant, complex colours to warn off their predators and have toxic secretions on their skins to fight those which are bold enough to attack them. These few examples show how predators and prey are locked in continuous struggles, from which the fitter ones survive and the balance of nature emerges.

The harsh wilderness is evident in other ways besides hunting and being hunted. Infectious diseases are caused by microbes, many of which are transmitted by animals. Other infectious diseases are carried by air, water or food. The pathogens range from viruses and bacteria to parasitic protozoa and worms of various kinds. The route of infection can be directly from the air we breathe, like the influenza virus. Other pathogens go through intricate life cycles, for example, schistosomes or blood flukes, which infect us through skin penetration when we take a dip in water where schistosome-infected snails live. More bizarre examples are found in the insect world, where, for example, a parasitic wasp lays an egg in a ladybird, which turns into a 'zombie', protecting the larva as it develops and eats the victim from inside. These examples illustrate the harshness of the wilderness in various ways and also show that, taken in the broad sense, it is not just located in wild forests or remote savannahs; it is in the midst of our civilization in the form of threats from infectious diseases and the fragile environment.

1.2 Wilderness Is Healing

The wilderness is not always harsh. Many species share a habitat without harming each other. Indeed, two or more species can thrive through helping one another. For example, *Rhizobium* bacteria live in root nodules of leguminous plants and help to fix nitrogen, that is, turns nitrogen from the atmosphere into nutrients for the plants, while it gains other nutrients and benefits from the plants—a mode of living together known as symbiosis. In some cases, the relationship becomes so close that the two organisms merge into one (the chloroplast, an essential component of plant cells responsible for photosynthesis, and the mitochondria, a component of both plant and animal cells responsible for generation and utilization of energy, were once microbes which became closely symbiotic with cells of early eukaryotic organisms, until they eventually merged together into single organisms) [2]. The chloroplast helps to build nutrients from carbon dioxide and water with sunlight as the energy source, and the mitochondria serves as the powerhouse of the cell by burning sugar to provide energy. Multitudes of bacteria on our skin and in our guts are also living in a symbiotic relationship with us. Among other functions, they help to digest and decompose our foodstuffs and help us to defend against unwanted infectious pathogens. In return, they receive food and habitation from us. Cultivation of a plant by humans or domestication of animals can also be considered as a form of symbiosis, albeit one planned by humans and not arising spontaneously by natural selection. From the wilderness therefore emerged agriculture and other forms of domestication as adoption of selective species for our benefit.

Despite the threats of the wilderness in various forms, it is also a source of our well-being, not only in a material, but also in a spiritual way. We often yearn to be back with nature, away from artificial surroundings of cities or villages. In the words of the author Henry David Thoreau, 'I believe in the forest, and in the meadow, and in the night in which the corn grows' [3]. Being back with nature refreshes our spirit, perhaps due to the fact that over the course of our long evolution, humans have only recently been divorced from nature to live in civilizations. Our spirit is therefore revived when we go back to our roots, even occasionally.

One of the great benefits of the wilderness is the fact that it is the source of various remedies for illnesses past and present, from the ancient practitioners of Chinese medicine to the traditional diviner-healers of Africa. Countless herbal products are included in the pharmacopoeia of traditional or alternative medicine. Many others are included in the list of modern drugs with proven efficacies. A few of them are:

- Artemisinin, from sweet wormwood (*Artemisia annua*), long used in China as an antihelmintic and now chemically processed to various derivatives with high efficacy against malaria
- Aspirin, from willow bark, prescribed for fever and aches since ancient times, now also used to reduce the risk of heart disease and stroke
- Caffeine, from coffee, tea and other plants, used as a stimulant by many cultures over the ages
- Digoxin, from the foxglove plant, traditionally used to treat various heart conditions but also known to have many adverse effects
- Morphine, a powerful analgesic and narcotic from the opium poppy, the use of which predates history
- Papain, a protein-digesting enzyme from the skin of the raw papaya fruit, used as a digestive aid

- Quinine, from the bark of the cinchona tree originally from Peru, introduced to Europe since the 17th century to treat malaria

---

Box 1.2 An Ancient Cure from a Common Plant [14]

Artemisinin is the name of a potent antimalarial drug which came from sweet wormwood (*A. annua*, *qinghao* in Chinese, see Fig. 1.2), a common plant used in China more than 2000 years ago against various illnesses. The name of the genus, *Artemisia*, is derived from Artemis, the Greek goddess of the wilderness, virginity, diseases and healing. The compound was identified in 1972 by Chinese medicinal chemists and has been used both as a drug and as a parent compound for many derivatives which are even more potent. Artemisinin derivatives are now used mainly in combination with other antimalarials in order both to increase the efficacy and to forestall the development of resistance. Nevertheless, resistance to artemisinin has been increasingly widespread, underlining the importance for developing new antimalarials.

Malaria is an ancient disease, caused mainly by two protozoan species, *Plasmodium falciparum* and *P. vivax*, and transmitted by the bite of the *Anopheles* mosquito. It is still responsible for close to 600,000 deaths per year, mainly in Africa. Like many other 'neglected diseases', it affects mainly the poor in tropical developing countries with poor health infrastructure. Quinine, another product from the wilderness, the cinchona tree, has been the main cure for the disease over many centuries, accompanied in the past few decades by such synthetic drugs as chloroquine and Fansidar (pyrimethamine–sulfadoxine combination), all of which now are threatened by emergence of resistance. The following chapters will give examples of efforts to deal with resistance and develop new effective antimalarials.

Other healing plants (Fig. 1.2) include the ebony tree, the fruits of which are traditionally used as antihelmintic (worm-killing), and *Senna*, the leaves of which are traditionally used as a laxative and for other purposes. Medically effective compounds from these plants are examples of materials, originally from the wilderness, long known and used as traditional remedies. Many of these traditional remedies have been investigated and have yielded effective molecules now used in modern medicine.

(Continued)

Box 1.2   (*Continued*)

(a)   (b)   (c)

Figure 1.2 Healers from the wilderness. (a) The ebony tree (*Diospyros mollis*), anthelminthic; (b) the Senna plant (*Senna alexandrina*), laxative; and (c) sweet wormwood (*Artemisia annua*), antimalarial. Pictures courtesy of BIOTEC Thailand.

Knowledge of uses for these and other agents has been accumulated over many years, indeed in many cases over millennia, through traditional wisdom passed through the generations [4–6]. The science of ethnopharmacy has developed around the use of traditional drugs by ethnic groups and integration with modern knowledge about drugs and their actions. It is very useful for an understanding of how local cultures utilize materials from their ecosystems to maintain health and fight illnesses. Moreover, it is a valuable source of information in the search for new drugs for global use, as history has shown. In the investigation of the use of traditional medicines as modern drugs, however, safety should be a main concern, since many of them may have toxic or unwanted side effects.

While some medicinal plants are now cultivated for our use, those which remain in the wilderness are also great potential resources. Many traditional herbal drugs contain ingredients from the wild, often because they cannot be domesticated easily or simply from expediency. Moderate harvesting of medicinal products from nature, as in the practice of traditional societies, does not disturb the balance of nature irreversibly. However, unlimited exploitation without replenishing the sources results in an unsustainable situation which jeopardizes the survival of the exploited species and threatens natural biodiversity. An example is licorice, an Asian plant used for pain relief as well as a flavouring agent, which is threatened from overharvesting and habitat destruction. Traditional belief in medicinal properties of such animal parts as rhinoceros horns and tiger bones, which has not been supported by scientific evidence, has also contributed seriously to the threats of extinction of these species.

1.3   We Are Living in a Molecular Wilderness

Apart from wild nature, which can be harsh or healing, there is another wilderness in which all of us live no matter where we are. Furthermore, this basic wilderness reflects the harsh and healing aspects of wild nature, and its understanding can point the way to our future development.

The wilderness is not necessarily displayed in wild surroundings full of untamed creatures. Nor is it an uncontrolled state of living, which most people have surpassed. There is a deeper roughness which still stays with us as the human society moves forward in its civilization, with all its technological progress. It comes from inside us and from our relationship with nature. Its roots are in the molecules of which we are made—it is the molecular wilderness.

As living beings, we respond to products of nature from consumption, contact or otherwise. We interact constantly with organisms around us. Some indeed come from distant wild areas, through their mobility or aided by wind and water, but most are common living beings with which we are familiar. Our interaction with nature, be it through products or organisms, is set by biological rules with molecular mechanisms at their roots. Importantly, the genetic traits of all organisms, including us, contain the molecular features which arose long before the rise of human civilization. The molecular basis of our life processes, as for all living beings, is the legacy of our past ecosystem. As the noted molecular biologist Jacques Monod said, '. . . every living being is also a fossil. Within it, all the way down to the microscopic structure of its proteins, it bears the traces if not the stigmata of its ancestry' [7]. Although we may seem to have control over our destiny in our controlled environment, it is still a haphazard world down there. No matter where we are, we are still living in a molecular wilderness, both through genetic determination and environmental conditions.

1.4 Chemicals as Universal Tools of the Wilderness

Living organisms interact with one another through various means, including sight, smell, sound and contact. Chemistry often underlies these forms of interaction, which include hostile attacks or defence actions and communication for various purposes, including feeding and reproduction. The vivid colours of butterflies which warn potential predators of their poor palatability, for example, come from chemicals which form the colour pigments and the light-deflecting properties of the nanostructures which make up wings. Skunks use sulphur-containing chemical agents from their anal scent glands to deter their enemies. In other cases, communication is restricted only to members of the same species, such as pheromones, chemicals given off by insects and various other organisms in tiny amounts to transmit alarm, indicate food sources or attract members of the opposite sex. Contact at the cellular level is also mediated by chemistry, through interaction between binding molecules and membrane receptors of the sense organs, for example. Such chemistry-mediated contact provides the basis for the senses, including sight, smell and taste. Chemical signals from pheromones are detected in many animals by special sense organs related to those for smell. In the case of hearing, sound vibrations stimulate an electrical response through opening of the auditory transducer channel with a mechanism yet to be revealed in molecular detail.

Chemical substances are major tools for attack, defence and repair, which are all vital functions for survival in the wild. The toxins in a snake bite, a bee sting or poisonous mushrooms in the tropical forest mentioned earlier are all chemicals made by different species to defend from, or attack against, their enemies. Snake venoms are mixtures of small and large bioactive molecules, ranging from amines and small peptides to various enzymes, which exert toxic effects on the circulatory, muscular and nervous systems. The principal component of bee venom is the peptide melittin, which exerts effects on many enzymes and physiological processes of the stung animal. Blister beetles contain cantharidin, a poisonous agent causing blistering, which has also been used for medical purposes since ancient times. Poisonous *Amanita* mushrooms, which resemble paddy straw mushrooms, contain alpha-amanitin, which causes liver damage, and phallotoxin, which causes intestinal upset.

We have learnt to use the chemicals from nature to our advantage, for example, those which microorganisms produce to antagonize or kill other microorganisms. Such chemicals are called antibiotics, the first example of which is penicillin, discovered by Alexander Fleming. With advances in medicinal chemistry, many antimicrobial compounds have been synthesized which are based on the structures of natural compounds. Other microorganisms called probiotic organisms, including some lactic acid bacteria, found in yogurt, for example, are thought to be beneficial to the host organisms, although the mechanisms of their actions are less clear.

When animals, plants or microorganisms are harmed, they often rely on the production of chemical substances to repair the damage. In higher animals these range from molecules which help to stop further damage like bleeding to those which help to form scars or regenerate lost tissues. Simple animals like *Planaria* and salamanders have even greater power to regenerate lost or damaged parts through mobilization of stem cells, in processes which require mediation of signal molecules.

Many animals have immune systems which have evolved to protect them from harm from other species or the environment. Immune systems comprise cells and molecules which have subtle specialized functions and work together in a complex manner to form robust protection mechanisms. The molecular basis of immunity involves detection and binding of antigens, or foreign agents, by antibodies. The cellular branch of immunity is activated to capture and destroy the intruding agents and further sensitize the whole immune system against the now-recognized intruders. These processes are mediated or modulated by small protein molecules known as cytokines. A number of chemicals from nature also have immunomodulating properties, that is, they modulate the functions of various immune components, either as suppressants or as enhancers.

Why do some chemicals from nature have the ability to act as poisons, drugs or modulators of our defence mechanisms? They are likely the products of evolution which help each species to survive and interact with the environment, including coexistence with other species, such as natural enemies and partners in symbiosis.

Since we share many similar characteristics with other animals, the actions of chemicals which have evolved for the purpose of survival of other animals sometimes, but not always, also apply to us. In many cases, chemicals evolved by plants to fend off their enemies, or for other purposes, also exert biological effects on us. However, we cannot predict which effects will be exerted, and we need to study them in order to be able to use them for healing and other purposes.

1.5 Interaction between Molecules as the Essence of the Wilderness

We can look upon the wilderness as an ecosystem in its natural state, with interactions among different organisms and between the organisms and their environment in general. In addition to using chemicals to control their own life processes, all living things make other chemicals to act upon other organisms for various purposes ranging from killing to finding sexual partners. These biologically active chemicals range in complexities, from simple molecules like nitric oxide, with just two atoms, to very large ones such as bacterial toxins with molecular weights of more than 300,000 daltons. In general, since many of these chemicals have to be transmitted from one organism to another, often through air or water, they tend to be of moderate size, with molecular weights of say less than 1,000. They can, however, have very complex molecular structures, the results of synthetic schemes peculiar to the organisms in order to achieve the biological purposes for which they have evolved to perform.

How do these chemicals exert their action? Here is where chemical processes of biological interaction differ mainly from those in a laboratory test tube. Instead of a chemical reaction arising from a simple encounter between two molecules, a biological effect exerted by a chemical from nature often results from a series of events triggered by specific binding with its 'target'. The target is a biomolecule, often a protein, or a large complex structure of many biomolecules. The exquisitely specific binding between the chemical and its target occurs through a complementary fit, mediated mostly by weak but multiple interactions between the various parts, akin to Gulliver being tied down by the tiny people of Lilliput.

Box 1.3 The Key to Unlock the Wilderness

At Robert Koch Platz in Berlin (Fig. 1.3) stand statues of two scientific and medical giants of the late 19th and early 20th centuries. Robert Koch, awarded the Nobel Prize in Physiology or Medicine in 1905, was famous both for isolating a number of important pathogens and for establishing 'Koch's postulates' which are still in use today for determining an organism as the cause of a disease [15]. In brief, the postulates state that the causal organism must be found in all cases of the disease examined, while absent in healthy hosts, should be isolable and able to be maintained in culture and should be able to produce the same infection in a healthy host.

Apart from pioneering work in sugars and purines, Emil Fischer, awarded the Nobel Prize in Chemistry in 1902 [16], made great contributions to the study of proteins, including enzymes. He made a postulate that an enzyme, the agent which catalyses the reaction of its substrate, is similar to a lock, while the substrate is similar to a key. A complementary fit between an enzyme and its substrate is like the fit between a lock and a key. Only the correct substrate 'key' can fit the enzyme 'lock'; other substrates will not fit the lock and will not be converted to something else. This postulate is still used today to explain the specificity and efficiency of enzymes, although it is now recognized that an enzyme may be flexible in nature and in many cases can change its configuration to fit the substrate.

The fit between the enzyme and the substrate is now recognized as one example of molecular recognition between biological molecules, which underlies the principle of many processes ranging from antibody–antigen interaction, cell–cell interaction and interaction of cellular receptors with a variety of molecules, including hormones and drugs. In the case of drug–receptor interactions, the receptor or drug target may lose its normal function in interaction with natural biological molecules or may sometimes be activated or changed in some other ways.

For our purpose in understanding the molecular wilderness, we should recognize that while Koch's postulates help to explain the wilderness at the organism level, Fischer's lock-and-key theory helps to explain it at the molecular level.

(Continued)

Box 1.3   (*Continued*)

Robert Koch
1843-1910

Robert Koch Platz, Berlin

Emil Fischer
1852-1919

Figure 1.3   Emil Fischer and Robert Koch.

These specific interactions between molecules, which might be called molecular recognition, and the following series of events indeed occur routinely in normal life processes of an organism and not only in interaction with other organisms. They arise from evolution of both the targets and the molecules which interact with them. A target is typically a protein, often with a nonprotein part attached, with a binding site composed of amino acids spatially arranged in such a way as to interact specifically with various parts of the binding molecule. The close fit and the ensuing process after the binding occur, not by design, but by the long process of natural selection from a pool of random genetic changes, which leads to natural solutions which work for the specific task required, for example, digestion, where an enzyme breaks down another protein molecule; neurotransmission, where a signal molecule is recognized by a receptor; and antigen capture by an antibody.

When a foreign molecule other than those normally encountered comes into play, the molecular target, the cell which harbours it and the whole organism can respond quite differently, resulting in various effects different from usual. An example of a harsh event in wild nature—damage from the bite of a poisonous snake—is due to the toxins exerting their effects on the targets in our blood and nerve cells. On the other hand, aspirin (a healing principle from nature) exerts its action through its interaction and subsequent inhibition of an enzyme which produces pain substances. Knowledge about how a molecular target responds to the molecules which interact with it (which ligands) can help us in the search for new drugs.

1.6 Tapping the Molecular Wilderness for Drugs

In addition to traditional knowledge about drugs from nature, searches for chemicals from nature have yielded many more which have proved to be good drugs with the ability to fight infections or cure other illnesses [6, 8, 9]. These searches might be said to constitute the process of 'tapping molecular wilderness'. Systematic searching for new agents from nature as potential drugs is an arduous task, and verification of the efficacy of traditional agents requires scientific studies of their effects at molecular, cellular, tissue and organ levels, culminating with studies on whole animals and humans. Normally, extracts of herbs or other natural sources are analysed and separated into various chemical components which are structurally identified and examined individually. In some cases, such studies have yielded compounds with good efficacies. More often, studies of the compounds with related structures, both from nature and synthesized in the laboratory, provide information about the parts of chemical structures or molecular features—called pharmacophores—which are important for their drug properties.

To proceed further to find the most potent drugs, chemical interactions of the pharmacophores and other parts of the molecules with the biological target must be examined. Just as in the case of normal life processes, the biological effects are triggered by molecular recognition of the ligand which carries the pharmacophore by the target, leading to further events such as inhibition or initiation of a chemical reaction, electrical responses or changes in cell membrane leakiness. In cases where the molecular target is known, an enzyme, for example, the effects of the compounds can be examined by inhibition of its normal function. In other cases, where the molecular target is not known, or where there are many molecular targets, the effects of the compounds can be examined at the cellular and physiological levels, with the end point as a measurable change in cellular properties or simply cell death. However, the underlying assumption is still that there are molecular recognition processes which lead to cellular and other responses.

From this initial discovery phase, we may be fortunate enough to find a set of compounds which can hit the target. However, to be useful as drugs, these molecules must be examined for their safety, or lack of toxic or serious side effects. Moreover, these molecules must have properties which will make them good drugs, including their ability to reach the target organs or tissues in good time, their stability and their conversion to other compounds in the body and the effect of these other compounds which must also be shown to be safe.

1.7 Tapping the Molecular Wilderness for Drug Targets

From the examples before, we can see that the process of tapping the molecular wilderness for remedies to our illnesses is more complicated than simply searching for molecules from nature which can act as drugs or promoters of immunity. We must also deal with how molecules act through specific interactions with their targets, and the ensuing effects. In the molecular jungle, a large variety of ligand molecules arose from coevolution with their natural targets so as to provide the mechanisms for life processes based on specific recognition. These processes include those within the organisms themselves and those underlying interactions between different organisms, such as in defence, attack and symbiosis. In the search for ligand molecules which can serve as useful drugs, for example, we should know the structure and function of both the ligands and their specific targets [10, 11]. Furthermore, we should know how the two partners interact in molecular detail so as to understand and be able to modify the ligand structure for our purpose.

We need to know the molecular targets for our potential drugs for another important reason. Biological molecules such as proteins, which form the majority of molecular targets, evolve by random mutations, giving rise to new structures with different recognition profiles. For example, use of an antibiotic to kill an infectious pathogen will naturally lead to emergence of resistant strains of the pathogen. The resistant pathogen can arise through various mechanisms, in which a common and important one is the change in structure of the original molecular target through mutation so as to avoid interaction with the drug, hence enabling the organism to survive. To be able to find the best drugs, and to change to new ones appropriately, it is therefore important to know not only the molecular features of the targets but also the mechanisms by which they can reduce the effectiveness and become resistant to the original drugs. Once this is known, it may be possible to develop new drugs which can deal effectively with the resistant pathogen, for example, by modifying the structure so as to enable them to recognize the changed target or to inhibit processes which gave rise to resistance to the original drugs.

It is important to realize that drug resistance does not arise only from mutational changes in the molecular target so as to reduce the affinity for the drug but also from various other mechanisms. In some cases there can be overproduction of the target so that there is sufficient uninhibited target to keep the pathogen or the cancer cell working. In other cases, the cell can bypass use of its target through an alternative route or develop mechanisms to destroy or reject entry of the drug. These and other mechanisms mean that studying structural change in the target alone is not sufficient, but we must also investigate the nature of resistance more widely, including search for ways to block its occurrence or for other effective targets.

1.8   General Outline of This Book

This is a book for general readers with some background in science, concerning the search for drugs, starting from molecular diversity found in nature, which might be called 'molecular wilderness'. The drug molecules from nature may be used as such or may be used as starting points for improved synthetic or semisynthetic drugs obtained from the interface of chemistry and biology. In some cases, the pharmacophores—essential molecular features for the drug properties—from natural molecules may be identified and modified to more effective ones. In other cases, nature provides the targets, such as essential enzymes from infectious microorganisms, from which synthetic drugs can be designed. The mechanisms of action of drugs can be discerned from studying the target–drug interactions. Nature may fight back, as when microorganisms become resistant to drugs, but we can again use the chemistry–biology interface to obtain drugs which overcome the resistance. The battle goes on, hopefully with victory for both humans and the balance of nature.

The book is divided into six chapters. After this first introductory chapter, the second chapter, 'Taking from Molecular Wilderness', examines how humans have over the ages discovered useful remedies from herbs and other natural substances, the nature of most of which become known only much later. Chemistry interfaces with biology to refine and produce drug substances from nature. The third chapter, 'Drug Templates from Molecular Wilderness', shows how original molecules from nature provide templates which can be modified to make better drugs. The fourth chapter, 'Drug Targets from Molecular Wilderness', discusses how molecular targets for drugs can be identified from infectious pathogens and cells which carry the targets and studied to provide opportunities for drug design and development. In the fifth chapter, 'The Wilderness Fights Back', it is shown how drug resistance and emerging diseases necessitate evolving strategies. The final chapter, 'Living with Molecular Wilderness', offers the guideline that coexistence and sustainability of nature should be preferred over overexploitation and subjugation of the molecular wilderness.

For convenience of readers not familiar with technical terms, a glossary is given at the back of the book. Other technical terms are explained briefly in the text on their first mention.

Chapter 2

Gifts from Molecular Wilderness

The art of healing comes from nature and not from the physician.
Therefore, the physician must start from nature with an open mind.
—Paracelsus, 1493–1541

Summary

Humans have over the ages discovered useful remedies from herbs
and other natural substances, the nature of most of which is learnt
of only much later. Chemistry interfaces with biology to refine
and produce drug substances from nature, often with the help of
knowledge from traditional medicine, which is still the major source
of provision of health care for most people in developing countries.
Tapping the molecular wilderness needs to be done sustainably and
responsibly, with fair benefit to the indigenous people whom we
must respect for their heritage and traditional wisdom.

2.1 Traditional Medicine: From Past to Present

The molecular wilderness in nature has been long tapped by
traditional medicine, even before the concept of the molecule was
born.

Tapping Molecular Wilderness: Drugs from Chemistry–Biology–Biodiversity Interface
Yongyuth Yuthavong
Copyright © 2016 Pan Stanford Publishing Pte. Ltd.
ISBN 978-981-4613-59-0 (Hardcover), 978-981-4316-60-6 (eBook)
www.panstanford.com

Human societies have relied on remedies from herbs and other natural substances to treat various ailments from fever to madness since ancient times. Typically, treatments were administered by respected healers, usually believed to have magical powers vested in them by the holy spirits. Many illnesses were believed to be caused by evil spirits, which needed to be purged by ritual means, of which the herbal remedies are only a part. Other treatments include surgical procedures and physical therapies such as massage and exercise.

With the rise of great civilizations, the art of medicine developed with different underlying principles and characteristics. Ayurvedic medicine, practised in India and nearby regions for thousands of years, has its roots in the Hindu and Buddhist religions. It stresses the importance of balancing mind and body among the essences of which the body is made (earth, air, fire and water). Diseases are believed to be caused by disturbance of the balance, for example, from overindulgence in pursuing pleasure of the senses, inappropriate behaviour and inappropriate lifestyle, and from passage of time or seasonal changes. Treatment of disease is aimed at restoring the balance and relies on the use of many herbs and spices, often in combination with physical and mental therapies.

Chinese traditional medicine is based on the principle of maintaining or restoring the balance between two opposites, yin and yang, which are present in various organs of the body as well as in other materials in nature. The former represents a cold, dark and quiescent state, while the latter represents a hot, bright and active state. Disease is attributed to excess or deficiency of one or the other. The body is also considered to have both tangible and intangible essences, loss in harmony of which leads to disease. Treatment includes use of herbal medicine and other practices such as acupuncture and moxibustion (herb-burning).

In the West, the ancient Greeks and Romans relied on the cultivation and use of herbal drugs for many ailments, as evidenced, for example, by the formulas prescribed by the famous physician Galen (AD 130–200). The Greco-Roman expertise was preserved and expanded by the Arabs and Persians, exemplified by the polymath Avicenna (AD 980–1037), who also developed their own sources, as well as taking materials from China and India. In addition to its contribution to traditional medicine in the West and the Middle East, this stock of early knowledge also led to the development of modern medicine and science.

Although traditional medicine practices in different parts of the world have different characteristics, they often have common aspects in using the holistic approach, considering the human as a part of the natural, social and spiritual environment. Disease is considered a result of disharmony between the patient and the environment, which includes plants, animals, objects, the landscape and spirits. Another common aspect is reliance on natural substances from plants, animals and minerals as a major part of the therapy.

In most countries, especially those with inadequate health services, traditional medicine continues to be practised alongside modern medicine. Herbal pharmacopoeia, collections of directions for identification of samples and preparation of drugs from herbs, have been compiled by many governments and pharmaceutical societies. Furthermore, many countries have included herbal drugs in the lists of officially approved essential drugs widely accessible to the public.

The continuing practice of traditional medicine offers alternatives for the public in the choice of health care. It also helps many countries in lessening the burden of providing modern medical services, especially in remote or poverty-stricken areas. The World Health Organization [17] estimated that more than 80% of people in many Asian and African countries rely on herbal medicines for at least some parts of their health care. Many countries have traditional medicine as a discipline, not only for vocational courses, but also as serious study for university degrees. They also recognize traditional medicine as a legitimate approach to health care. Thailand, for example, has a government hospital dedicated to traditional medical treatment using herbal products and other approaches such as massage. This is a continued tradition of practice handed down over the centuries (see Box 2.1).

Box 2.1 Traditional Medicine as a Heritage

Situated next to the Grand Palace in Bangkok, Wat Po (Fig. 2.1), is an ancient beautiful monastery with many distinctions. The most important is that it was the first 'open medical school', a place for

(*Continued*)

Box 2.1 *(Continued)*

learning, especially in the arts of healing. King Rama III ordered a major restoration of the monastery almost 200 years ago, in which he made it into an open learning place. As ordered by the king, around the pagodas and in the open halls are inscriptions of knowledge of traditional medical practice handed down over the centuries, including instructions for making herbal drugs for various ailments, for massage and physical exercises as taught by the holy men (rishis). Furthermore, the monastery grounds contain many herbal plants specified in the inscriptions—an open laboratory. To this day, the monastery continues to be the training ground for traditional medical practitioners, who can choose various courses ranging from contortion exercises (rishi dud ton) to herbal massage to her]bal medicine.

Traditional medicine continues its popularity among the general population. For example, Abhibhubejhr (read apai-pu-bate) Hospital in the east of Thailand is a famous place for health care through 'good traditional medicine practice'. It makes its own herbal products using GMPs, using raw materials bought from trained farmers around the area. Significantly, its products are mostly herbal formulations well known since ancient times and include foods, cosmetics and other traditional medical materials, mostly suited for mild ailments which can be self-treated without consulting a doctor, a preferred course of action for most people. Examples include capsicum cream, containing capsaicin, the hotness ingredient of chili pepper, for treatment of muscle pain; fatalaijone capsules, containing andrographolide from the extract of *Andrographis paniculata*, for treatment of diarrhoea, fever and sore throat; and turmeric capsules, containing curcumin from the extract of *Curcuma longa* rhizomes, for treatment of flatulence.

Figure 2.1   (a) Wat Po, the first 'open medical school' (from Wikimedia Commons), (b) a rishi in exercise and (c) some herbal ingredients as used by Abhibhubejhr (from www.abhiherb.com).

However, risks in the use of therapies not supported by evidence, poor standards of practice or even unethical practice by charlatans must be taken into account and appropriate measures taken, including regulatory measures and public information. Some remedies are harmful, such as traditionally prepared drugs containing heavy metals or highly toxic compounds. For example, *Aristolochia* plants widely used in China and elsewhere for various ailments contain aristolochic acid associated with chronic kidney disease and cancer [18, 19]. Worse still, adulteration of traditional medicines with pharmaceutical agents, such as steroid drugs, can have harmful or even fatal effects. It is therefore important to have good policy and practice of traditional medicine to serve and protect the public.

2.2 The Value of Traditional Wisdom and the Importance of Validation

Although traditional medicine has been widely used in various societies around the world for health care of people, especially those who do not have access to modern medical practice, its potential in contributing to modern medical practice has still not been fully realized. One reason is the difference in the systems of knowledge, which have unique characteristics and are not easily connected to concepts in modern medicine. Another is the variability of the practices and materials used by traditional healers, often regarded as having connection with magic or mysticism. Yet another is the lack of verifiable information often regarded as secret by the healers. Nevertheless, much useful information has been collected from sources all over the world and compiled into a body of knowledge known as ethnomedicine.

Ethnomedicine may be broadly defined as the practice of healing sickness and maintaining health, often involving the use of materials from nature as medicine according to traditional beliefs. The majority of the materials are herbal in nature, and ethnobotany, or knowledge about plants traditionally used as medicines, is therefore a major contributor to ethnomedicine. Another contributing subject, ethnopharmacology, can be defined as the study of the biological activities of herbal drugs on humans, including experimental investigations on animals and other systems in order to discover the active ingredients and their properties, hence leading to the discovery of new drugs.

Discovery of new drugs for use in modern medicine is a major contribution from ethnomedicine. It has been estimated that the majority of drugs isolated from plants have ethnomedical use: the plants from which they were derived have been used in traditional medicine in various cultures. Important though the contribution of ethnomedicine is to modern drug discovery, no less important is its value in strengthening good practice of traditional medicine, which is still a major part of the public health picture in most developing countries. In the compilation of traditional beliefs in the healing power of materials from nature, ethnomedicine allows extensive comparison across various cultures and enables reasonable tentative conclusions about the validity of the traditional materials used. These tentative conclusions should lead to scientific investigation into the roots of such beliefs, with eventual validation of their value, or in many cases also of the dangers to be avoided.

2.3 Critical Issues on Drugs from Traditional Medicine

As traditional medicine is still a major health care tool for many cultures, especially among people with moderate means of livelihood, validation of its value is very important in preservation of their beliefs and ways of life. However, it is also very important to ensure the safety of traditional medical practice and avoid harm from charlatans or from honest beliefs which turn out to be wrong or do more harm than good. Modern medical science has a large role to play in both the validation of the efficacy of herbal and other materials used in traditional medicine and the assurance of quality and safety of the materials and the practice as a whole. Rather than separating traditional from modern medicine, it is better to try to integrate the two systems and gain the best of both. While traditional medicine can gain from passing the rigorous tests of modern medical science in safety and efficacy, it can contribute knowledge in the use of materials from nature selected from many generations of practice, from which active molecules can be identified and developed further as modern drugs.

Although traditional medicines in various parts of the world are based on different principles from modern science, there are many areas where integration is possible and beneficial. For example, the variability of different medical preparations can be monitored by modern analytical techniques, enabling standardization and quality control. The analytical techniques can also detect dangerous additive substances which are added by unscrupulous practitioners for increasing the apparent efficacy of treatment but with untested, unknown or unacceptably harmful consequences. The interaction of various components in the preparations, which often give rise to synergistic effects, can also be studied and better understood by current pharmacology.

There are nevertheless many critical issues in the integration of traditional medical practice with current modern practice. For example, different concepts of wellness and illness can lead to different interpretations of the clinical outcome of intervention by traditional medicine. Furthermore, other approaches are often used in combination with herbal drugs, such as acupuncture and meditation. The fact that herbal drugs are often made from a combination of various ingredients from nature makes variability an inherent property of these preparations. Although many countries, especially developing countries where the majority of the population is poor and dependent on traditional medicine for health care, have adopted herbal drugs in their official pharmacopoeia, various issues still remain concerning regulation of traditional medicine. Indeed these critical issues are present in all parts of the world, even in the developed countries, as many seek alternative treatments for their health problems. In view of the social importance of traditional medicine, efforts to solve these difficult issues must continue.

2.4 Biodiversity as a Source of Drugs from Nature

Drugs from traditional medicine come mostly from plant products. However, vast as this source of drugs from traditional wisdom is, it is still small compared to what nature has to offer from its biodiversity [9, 20]. It has been estimated that there are some 250,000 higher plant species, of which only a small fraction has been either used in traditional medicine or explored by modern methods as sources of useful drugs. For example, it has been estimated [6] that only 6% of the higher plants have been screened for biological activity. Only just over 100 compounds of defined structures from 94 species have been identified from these plants, which are used globally as drugs. Significantly, 80% of these are used in ethnomedicine in some form. They include, in addition to the partial list we saw in Chapter 1, atropine (poison antidote, arrhythmia and other treatments, from the deadly nightshade, *Atropa belladonna*), ephedrine (stimulant, from *Ephedra sinica* used in the Chinese herb *ma huang*), gossypol (male contraceptive, from the cotton plant, *Gossypium*), ouabain (heart stimulant, originally used as arrow poison, from *Strophantus gratus*) and stevioside (sweetener, from the sweet grass *Stevia*) [6, 21]. Therefore, we can conclude that there is still a wide scope for exploration of plants for new drugs, and information from ethnomedicine is invaluable in such exploration. Drugs derived from plants without ethnomedical information include L-DOPA (psychoactive drug used for Parkinson's disease, from the velvet bean, *Mucuna pruriens*), taxol (anticancer drug, from the Pacific yew tree, *Taxus brevifolia*) and vincristine (anticancer drug, from the Madagascar periwinkle, *Catharanthus roseus*).

Rich though the higher plants and animals are as potential sources of drugs, they are just a part of biodiversity which can be tapped. Animals, including many insects, fishes and amphibians, also provide many biologically active compounds, some of which are poisons, while others can be used as drugs. Insulin, the drug for diabetes, was derived from the pancreas of pigs and cows before its production by recombinant DNA technology. Calcitonin hormones used to treat osteoporosis were derived from salmon before modern manufacturing by synthesis or recombinant DNA technology was available. Captopril, used for treatment of hypertension, is a peptide which was first found from the venom of the lancehead viper and used as an arrowhead poison by a Brazilian tribe. Poisonous toxins from the skins of many vivid-coloured frogs from South America, which are produced as a defence against predators and used by indigenous tribes as dart poison, have pain-killing, muscle-relaxing and other medical values still to be fully investigated. However, continuing traditional beliefs in healing values of such animal products as rhinoceros horns or tiger bones have no scientific basis and are sadly threatening the existence of these species.

The richest sources of drugs from nature, however, are microorganisms, which range from bacteria through algae and fungi through small primitive plants. Microbes produce a wide variety of antibiotics as weapons to kill or harm other microbes and win an advantage, for example, in access to food resources. In addition to antimicrobial drugs, microbes have also provided many important drugs such as statins, for lowering cholesterol (see Box 2.2), and

---

Box 2.2 Statins: From Nature to Every Medicine Cabinet

Akira Endo (born 1933) [32] dreamed of becoming a scientist from the age of eight. He had an early interest in fungi and as a biochemist later developed an interest in cholesterol biosynthesis. While working in New York, he noticed the rich dietary habits of Americans compared with the Japanese, and a large number of overweight people with coronary heart problems. Although at that time, around the 1960s, there were some lipid-lowering agents, none were very safe or effective. It was then known that humans produced more cholesterol than that obtained from diet, through what is called the mevalonate pathway, and the rate-controlling enzyme in this process is 3-hydroxy-3-methylglutaryl-coenzyme A reductase (HMG-CoA reductase), which converts HMG-CoA into mevalonate. Endo's experience in microbial metabolism led him to hypothesize that microbes would produce inhibitors of this enzyme to interfere with sterol synthesis in other microbes.

A search of 6000 strains of microbes identified 2 fungi with such predicted activities. Further work showed that one of these, *Penicillium citrinum*, produced a compound later called compactin, which is a potent inhibitor of HMG-CoA reductase. Interestingly, the structure of compactin has a part which is very similar to mevalonate (Fig. 2.2). As shown subsequently, it inhibits the enzyme by binding to the active site, blocking the substrate and the reaction of the enzyme. Subsequently, other compounds were identified from other microorganisms with similar biological properties. Treatment of patients with high cholesterol with compactin and similar compounds led to significant lowering of cholesterol in their blood, with good clinical outcome. Presently, this group of compounds, known as statins, is generally accepted as being highly effective for lowering cholesterol levels and reducing the risk of heart disease and other related conditions.

(Continued)

Box 2.2 *(Continued)*

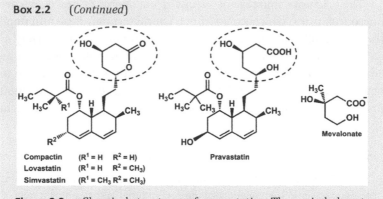

Compactin   ($R^1 = H$ $R^2 = H$)
Lovastatin   ($R^1 = H$ $R^2 = CH_3$)
Simvastatin   ($R^1 = CH_3$ $R^2 = CH_3$)

Pravastatin

Mevalonate

Figure 2.2   Chemical structures of some statins. The encircled parts are similar to mevalonate and bind to the active sites of, and inhibit, the enzyme HMG-CoA reductase, which catalyses the production of mevalonate from HMG-CoA.

cyclosporine, for suppressing the immune response. It is not known how many species of bacteria exist: estimates range from many millions to 10 billion. Many of these microorganisms live in our surroundings, in the soil, the ponds and the air. Even each healthy individual harbours more than 10,000 species of microorganisms on his or her body, including the gut, the skin and the nose. Others, called extremophiles, live in extreme environments such as hot springs and deep oceans. They provide a rich collection of diverse compounds which still wait to be identified and used as drugs and other useful agents.

The oceans are rich sources of biodiversity, covering more than 70% of the earth's surface. Although they have not been much used in traditional medicine because of relative difficulty in obtaining the materials and in culturing the microorganisms or farming the animals and plants, these marine sources have provided many important bioactive agents, some of which have been developed as drugs and cosmetics or are in clinical trials. Sponges, tunicates, corals and molluscs are some of the marine organisms from which compounds of importance to human health have been identified and developed as drugs [22]. An example is the group of toxins known as conotoxins from cone snails found in tropical seas, which are used as analgesics for chronic pain [23]. Another is trabectedin from sea squirts, which is used as an anticancer drug. Other compounds, such as spongothymidine and spongouridine from the sponge *Cryptotethia crypta* have good antiviral properties, and from these the related drugs Ara-A and Ara-C were developed. With new advances in culturing and farming of marine organisms, it is likely that many more bioactive molecules will be discovered in the near future, some of which will lead to new drugs.

2.5 Drug Discovery: From Biology to Medicine through Chemistry and Allied Sciences

Discovery of new drugs from plants, microorganisms and other living organisms now usually takes a multidisciplinary approach, often involving rapid random screening of samples for desired pharmacological and other properties, using a combination of biological and molecular techniques. Botanical samples are extracted with water or other solvents (hexane, ethanol, acetone, etc.), the choice of which depends on the type of natural products to be studied—polar compounds tend to be extractable with water and nonpolar compounds with nonpolar solvents. The extracts can be examined for the properties of interest, such as killing of microorganisms, or for such specialized functions as enzyme activities known to be associated with pharmacological properties. Positive samples can be analysed further for the active constituents, which can be identified chemically and their biological activities studied further. Plants already used in traditional medicine are normally preferred raw materials for investigation, as there is a greater than random chance of discovering active compounds. As a typical example, many groups have investigated the herbaceous plant *Andrographis paniculata*, which is commonly used to treat infection and fever in many cultures throughout Asia and the Middle East [24–26]. The ethanolic extract of whole or parts of the plant, followed by separation by chromatography into various fractions, yielded many compounds of various chemical types. Among these, andrographolide has been shown to be an active constituent with many pharmacological properties in line with traditional beliefs.

Investigation of active compounds from microorganisms usually starts with isolation and culture of various microorganisms present in the source materials, such as hot springs, sewage or soil samples. The culture conditions are normally set up to maximize the chance of finding the types of microorganisms of interest, such as the presence of special nutrients. Typically, a diluted suspension of the source of microorganisms is grown on an agar-containing culture medium, producing a number of colonies of microorganisms (bacteria, fungi, etc.). The antibiotic properties of the colonies, for example, can be tested by overlaying the culture with target pathogenic microorganisms. The colonies producing active antibiotics can be identified by the absence of growth of the target microorganisms and can be subject to large-scale culture in order to isolate and identify the active substances.

The method for obtaining antibiotics was pioneered by Selman Waksman, who codiscovered streptomycin and other antibiotics in the 1940s [27]. For over two decades, this approach proved very fruitful for finding antibiotics from natural products made by microorganisms. However, all antibiotics are effective only for a limited time span, since pathogens eventually acquire resistance through a variety of mechanisms, necessitating a search for new antibiotics. The search for new and different antibiotics gets increasingly more difficult, since empirical methods tend to turn up more of the same. We will return to this problem in Chapter 5.

The general method for discovery of antibiotics is akin to how Alexander Fleming discovered penicillin in 1928, a serendipitous observation that a contaminating mould produced a substance which killed bacteria in culture. In his humble way, Fleming taught us a great lesson when he said in his Nobel lecture [28] 'My only merit is that I did not neglect the observation and that I pursued the subject as a bacteriologist.' Great though this discovery was, it would not have advanced so much as to become the first great antibiotic without the contribution of Howard Florey and Ernst Chain, who demonstrated its therapeutic value and succeeded in making and refining it in sufficient amounts for human use, just in time to treat injured soldiers at the end of the Second World War. The chemical structure of the drug was determined by Edward Abraham, working in Oxford with Chain and Florey, and the three-dimensional structure was determined by Dorothy Hodgkin.

This exciting historical account demonstrates the general principle of drug development from natural sources, including plants, animals and microorganisms. Once biological activity of interest, such as an anti-infective property, is identified from an extract or directly from a microorganism, much work remains in order to find the identity of the active compound, or in many cases the mixture of compounds, and to progress further to turn each into drugs. Briefly, each compound from the starting mixture must be purified and subjected to various analyses, including elemental analysis (to determine the atomic ratio of components), mass spectrometry (for molecular weight and structural detail), infrared and ultraviolet spectroscopy (for molecular property), nuclear magnetic resonance spectroscopy (for molecular structure) and if the compound can be crystallized X-ray spectroscopy (for three-dimensional structure).

Not only must the structure be found, but also its synthesis should be achieved in the laboratory, and subsequently, if it proves to be of promising drug value, a scaled-up synthesis is needed for extensive testing. In cases where the compound cannot be synthesized chemically, a less preferable alternative of using the natural source must be followed by establishing optimal extraction and purification procedures. Synthetic versions are usually preferred since most bioactive compounds are present in tiny amounts in natural sources and so are more difficult and expensive to obtain from the latter.

Furthermore, as the next chapters will show, discovery of an active compound from nature is usually only the beginning of a long search for the optimal compounds to be developed as actual drugs. The compound presents a pharmacophore, or an active part of the molecule responsible for the pharmacological property, but this can be optimized, by design and synthesis of compounds of various structures with the same or similar pharmacophore, so as to have increased efficacy, lower toxicity and other properties required for a good drug. In many cases improved compounds can be designed from knowledge of interaction with their molecular targets or receptors in the microbes or in humans. Such investigation is in the realm of medicinal chemistry. This early stage of drug development, taking place at the interface of biology and chemistry, is diagrammatically shown in Fig. 2.3.

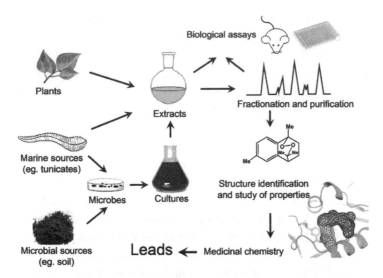

Figure 2.3   Early stage of development of drugs from natural products. Samples (plants, marine organisms, soil samples, etc.) containing active natural products are extracted or, for microbial samples, cultured and extracted. The extracts are fractionated and subject to biological assays for identification of active components. After purification, the molecular structure is identified and other properties of the active components investigated. This is usually followed by design and synthesis of related molecules and testing for optimization, resulting in drug leads for further development.

Drug development depends on many other allied sciences, many of which are interdisciplinary in nature. The fate of an administered drug can be studied through pharmacokinetics, which includes investigation of the extent and rate of absorption, distribution, metabolism and excretion (ADME). The biochemical and physiological effects of the drug on the various systems of the body must also be studied through the science of pharmacodynamics. The drug may exert stimulating or inhibitory effects on the various bodily systems through interaction with cellular receptors or other mechanisms. For anti-infective drugs, the effects on the various functions of the target pathogens must also be studied. These studies are parts of the broad subject of pharmacology, which also encompasses the areas of drug interactions, drug effects, including not only desirable but also toxic, mutagenic and other undesirable side effects, optimum dosage and routes of administration. Formulation of the drug is also an important step, where the active substance is combined with other ingredients in order to produce the final medicinal product which will give the optimal effect. The product can be in oral (tablet or capsule), injectable or other form according to information from comparison of effectiveness of various routes of administration. They have to be produced through good manufacturing practices (GMPs), which are regulated by drug-producing countries.

New drugs have to be tested for safety and efficacy, first through experimental animals and then through clinical trials on humans. Testing with experimental animals, called preclinical trials, is normally done with at least two species, say mice and dogs, but can also be done with other experimental animals, as appropriate. After a drug candidate passes these tests, application can be made to the regulatory authority for permission to undergo clinical trials. Clinical trials are done on humans and are divided into three phases before marketing, and a fourth, postmarketing, phase. The first phase is screening for safety, in which the drug is given to a small number of normal, healthy people (20–100) to determine the safe dose range and identify side effects. In the second phase, the drug is given to a larger group of people (up to several hundred) to evaluate the efficacy of its actual intended medicinal use and to further evaluate its safety. In the third phase, the drug is given to a large group of people (up to several thousand) to confirm its efficacy, to compare with current available treatments and to collect information for safety in its use. Drugs which have passed the three phases of trials successfully can be registered for use with the authorities of the countries where they will be marketed. The long process required for drug development and approval has a high failure risk, and those which are successful typically require some 10 years and research expenditure of several hundred million dollars. The fourth phase is done after marketing of the drug in order to obtain large-scale information on risks, benefits and optimal use. The whole cycle of drug development is shown in Fig. 2.4.

Safe and effective use of drugs in the market is the subject of pharmacy, a science which links chemistry with medicine, and includes dispensation of medications and collection and review of information on drug safety and efficacy. The subject of pharmacognosy, or the study of medicines derived from natural

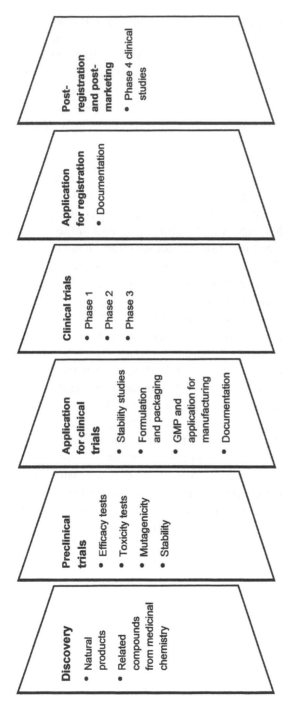

Figure 2.4 Cycle of drug development: The molecule discovered to have a good drug potential is subject to further testing, called preclinical trials, for efficacy and potential toxicity in various systems—molecular, cellular and whole animals. If it passes these tests, the drug can be approved for clinical trials in humans, which will require good manufacturing practices for making scaled-up quantities, with prior formulation and stability tests. The clinical studies are done in three phases, from which documentation on safety and efficacy is produced and submitted for registration of the drug. After approval, the drug can be produced and marketed and is subject to the fourth phase of clinical studies for further evaluation of the drug.

sources, is traditionally regarded as a branch of pharmacy. Finally, it should be noted that drugs are only a part of medical treatment, which comprise other procedures, including diagnosis, and other means of care for patients, as provided for by the medical profession. They should therefore be integrated with other elements of health care. Drugs from natural products are often used as nutritional supplements and are accessible to a large number of people, not only patients with specific diseases. The fact that products from traditional medicine have been used over the centuries does not mean that the safety and efficacy testing required for new drugs is not necessary, since the formulation of drugs, dosage and routes of administration is not the same. In some countries such as the United States, however, herbal products and traditional medicines are regulated together with dietary supplements as food and not as medicines.

2.6   Origins and Classes of Natural Products

Natural products include all substances produced by living organisms and as such are very diverse and difficult to classify in relation to their pharmacological properties [29]. However, we can understand the biological functions and drug-like activities of many natural products through an understanding of how and why they are produced in various organisms. This understanding comes through a study of metabolism, or the series of reactions in organisms leading to various products. Roughly, metabolism can be divided into two types, primary and secondary. Primary metabolism comprises the basic reactions producing the metabolites which are practically needed by all living systems. For example, the glycolytic pathway is a part of primary metabolism for producing various metabolites from glucose. The energy released by this process makes ATP, the basic energy currency, and NADH, the basic reducing power. Glycolysis occurs, with variations, in nearly all living organisms. Another example is the tricarboxylic acid cycle, or the Krebs cycle, which occurs in the mitochondria of aerobic organisms. In this essential process, acetate derived from carbohydrates and lipids is oxidized to generate energy with carbon dioxide as a by-product. Biosyntheses of proteins, nucleic acids, carbohydrates and lipids also generally follow similar mechanisms, with some variations in various organisms.

In contrast, secondary metabolism is the specialized series of reactions occurring in different fashions in specific organisms, resulting in specific compounds known as secondary metabolites. These compounds serve specialized purposes, for example, as toxic agents for the defence of the organisms against enemies, as colouring agents to warn potential enemies or as attractants towards the same or other species. Antibiotics are examples of secondary metabolites made by microorganisms to defend against other living organisms, which work by inhibiting their essential functions. Although there are overlaps between primary and secondary metabolism, in general the great variety of natural compounds representing the molecular wilderness are the products of secondary metabolism, although obviously an unclear borderline exists between these and the general molecules, like proteins, carbohydrates, lipids and nucleic acids, which are the basic ingredients of all living beings.

Secondary metabolites are generally formed from joining and arranging of building blocks obtained from primary metabolism. These building blocks can come most often in units of 1, 2 or 5 or less often with a different number of carbon atoms. The 1-C unit may be furnished by the amino acid methionine and the 2-C unit by acetate in the form of acetyl coenzyme A, while the 5-C unit (called isoprene) often comes from mevalonate, itself derived from acetyl coenzyme A. Other units of greater complexity, usually derived from amino acids, are also used, often with nitrogen attached. From these building blocks, the natural products are constructed with the help of specific enzymes into various types of compounds through a number of chemical reactions, ranging from combining together in various fashions to rearrangement and oxidation-reduction.

Because of the complexity of the natural products, they cannot be easily classified. Suffice it to say here that the main classes of natural products often encountered are:

- Polyketides. A large class of natural products from microorganisms, plants and animals, basically formed from coupling of 2-C (acetyl) units. Many polyketides have drug properties, including tetracyclines, macrolide and polyene antibiotics.

- Terpenoids and steroids. Two groups of structurally related compounds composed of isoprene (5-C) units joined in various ways (two isoprene units form a monoterpene). Many terpenoids are essential oils with strong tastes and smells. Many steroids are hormones.
- Phenylpropanoids and coumarins. Compounds with an aromatic ring and a 3-C side chain, originating from the shikimate pathway. Many are essential oils with strong tastes and smells.
- Flavonoids. Compounds resulting from a combination of shikimate and acetate pathways. Many are colour pigments found in plants.
- Alkaloids. Nitrogen-containing, mostly basic, compounds derived mostly from amino acids. Many have strong biological (drug, narcotic or toxic) activities. These include quinine, atropine, morphine and nicotine.
- Peptides, proteins and related derivatives. Peptides and proteins are normally made by ribosomes through genetic expression from RNA. Some nonribosomal peptides, many of which have antibiotic properties, are made by special enzymes of microorganisms. Antibiotics such as penicillin and cephalosporins are modified peptides.

2.7 Genes as Sources of Natural Products

The discovery of bioactive natural products from microorganisms normally depends on the ability to culture microorganisms at a scale in which it is possible to extract the products for further study and at an even larger scale when large quantities are needed for drug development. Often, the inability to culture microorganisms is the stumbling block. There are now general methods to grow microorganisms in a simulated natural environment rather than standard culture conditions, and therefore possibilities to discover new bioactive natural products. Even when the microorganisms cannot be cultured, promising new genomic technologies now allow production of a number of natural products, at least to a level where further studies are possible [30]. Metagenomics is the science of environmental genome exploration, in which the genetic information of all organisms in an environment is mined for genes encoding novel biosynthetic enzymes. This genetic information can be obtained from direct sequencing of the combined DNA from many different microorganisms in an environmental sample. By looking carefully along the DNA sequence (assisted by computers), genes can be isolated and transplanted into other cell types which can be cultured easily. From these surrogates, we can discover new biomolecules [31]. Metagenomics is also useful for the study and production of natural products from animals and plants. Some active compounds have been found in organisms such as sponges and sea squirts but in variable quantities. Often, it is found that these compounds are produced not by the animals themselves but by associated symbiotic microorganisms, usually not culturable. Metagenomics is a promising tool for study of nonculturable microorganisms and their products and for vastly increasing the scope of the molecular wilderness. It is being used to study the environmental genome of various niches, including the soil, air and water of various sources. Even the microorganisms in and on our body, the human microbiome, are being studied this way. This modern approach lends promise to the discovery of new knowledge and products useful to us, for example, products from useful microorganisms to help to keep us healthy or defend ourselves from harmful enemies.

Natural products are produced from microorganisms, plants and other sources through biosynthesis pathways, which are achieved by a series of enzymes. Advances in gene manipulation now allow us to add or remove genes for such biosynthesis, and therefore by making new pathways we can produce new compounds, so-called 'nonnatural' natural products. This subject will be explored further in Chapter 4.

Other genomic technologies allow production of natural products without depending on the plants or other living organisms which are the original sources of the products. Production of artemisinin and related antimalarials normally depends on growing and extracting the product from the plant *Artemisia annua*. Using genomic technology in inserting the genes from the plant into bacteria and yeast, it is now possible to produce artemisinic acid, the precursor of artemisinin from these microorganisms, and use further chemical synthetic methods to make the drugs. This is an example of synthetic biology, which in this case may also be called metabolic engineering, where products can be made from modified, or even wholly synthetic, organisms. Other natural products which may be produced by synthetic biology in the near future include saffron, the world's most expensive spice, and resveratrol, a compound found in red grapes used as a health supplement. Although a promising approach, this new way of tapping the molecular wilderness raises some ethical issues similar to the conventional way of extracting products from nature, the resolution of which would enhance its social value (see Box 2.3).

Box 2.3   Ethical and Social Considerations for Tapping the Molecular Wilderness

- Traditional medicine remains an important part of health care for most of the developing world, especially for those with moderate means of livelihood. To maintain or enhance its value, efforts must be made to ensure good practice. Science can help in confirming the validity of claims of efficacies of ingredients from nature, and in finding out the limits and potential dangers of their use. The benefit of following the wisdom of our ancestors in looking after our health must be weighed against the risk of poor practice, which might be ineffective, or even dangerous. Realising the value of heritage from the past, regulatory agencies, the civil society, social media and other concerned groups should build vigilance in the practice of traditional medicine with the help of modern science.

- Nature provides us with a large number of products which can be tapped for traditional and modern medical uses. In doing so, we should make sure of the sustainability of the natural sources. Local people who go to the forest to collect herbs can do so sustainably, since they take moderate amounts and give nature a chance to renew itself. Once this is done on a large scale by people who only take the products without making sure of sufficient regeneration, biodiversity is threatened and the outlook for future generations is dim. Many herbal products come from plants which can be grown and harvested by farming, a suitable method for ensuring sustainability. Green production should be the standard practice, ensuring that biodiversity is not threatened in tapping the molecular wilderness. With animals, there are also other considerations, such as avoidance of cruelty. A large part of illegal trade in animals or their parts is unfortunately due to demand as ingredients of traditional medicine. Science can help in the efforts

(Continued)

Box 2.3 (*Continued*)

to conserve biodiversity, for example, by offering alternatives to obtaining products from nature by chemical synthesis or use of recombinant DNA technology.

- Ethical and social concerns have been raised by the practice of bioprospecting (the survey and use of materials from a biodiverse environment mostly located in tropical developing countries) and biopiracy (the illegitimate appropriation of materials and intellectual property of the communities). In the latter case, controversies have arisen when researchers or corporate organizations have obtained patents on products and processes which originated from indigenous communities, and were then developed further by modern research. Sensitized by such concerns, voiced in the 1992 Earth Summit in Rio de Janeiro, an international agreement to combat biopiracy and share benefits from research on natural resources was concluded in Nagoya in 2010. Under this agreement, called the Nagoya Protocol on Access to Genetic Resources and the Fair and Equitable Sharing of Benefits arising from their Utilization (ABS), those seeking to use genetic resources or traditional knowledge for research or commercial purposes must obtain approval from both the government and the indigenous communities and agree to benefit-sharing, including financial, intellectual property rights and other benefits. Many countries and stakeholders have responded positively to this agreement. The European Commission, for example, has introduced a regulation to implement this protocol in the European Union (EU) and plans associated actions, such as compiling a register of trusted seed banks, botanical and other collections, for responsible supply of genetic resource samples.

- The alternatives to harvesting from nature, using genomic and chemical technologies, to produce or improve upon the products of nature are beneficial, as seen in this and the following chapters. However, such alternatives carry some ethical implications. In addition to the problems of ensuring the safety and efficacy of the processes and products, deeper questions need to be asked of the new technologies. In making new forms of life, with many new, and in some cases entirely artificial, genetic elements, we should be careful not to go down an increasingly slippery road. Controversies over genetically modified organisms (GMOs) in the past (mostly organisms with just one or a couple of foreign genes) have been long and intense. We should learn the lessons and move forward carefully with balanced ethical and social considerations.

2.8   Tapping Molecular Wilderness Sustainably

Biodiversity offers a 'medicine cabinet', from which we can discover a number of drugs. Many therapies were derived from knowledge handed down through countless generations, and these still serve a large number of people, especially in developing countries, in the form of traditional medicine. Other drugs were rediscovered as active components of traditional medical formulae. Yet others were discovered anew from investigation of plants through random screening or otherwise. Some were discovered and developed just in time to save us from new problems, for example, antimicrobial drugs against microbes resistant to existing drugs.

Tapping the molecular wilderness involves discovery and use of molecules from nature for the purpose of curing diseases, maintaining health and other benefits. It begins with what is called bioprospecting, or searching for useful molecules in a biodiverse environment, often armed with prior knowledge from ethnomedical or folklore information. When this is done by the local populations in their own environment, this activity poses little threat to biodiversity, which can spring back quickly. However, the real threat occurs when bioprospecting is done on a large scale, looking for raw materials from nature with the intention of extracting them for profit without means of ensuring sustainability, often by people from outside the communities. Such a practice is not sustainable. The threat is compounded when indigenous knowledge is used by outsiders for profit, both in making products and in claiming intellectual property rights, without or with only little compensation for the indigenous people, a practice known as biopiracy. The local authorities are usually too weak, or too prone to outside influence, to exert a counterbalancing force to the practice. Admittedly, outsiders, including researchers and industries, have a legitimate reason to explore biodiversity, which can be considered the heritage of humankind. It is therefore up to all the stakeholders, including the indigenous communities with materials and intellectual heritage at their disposal, their governments, outside researchers, public and private sectors of developed countries and civil society bodies, to act jointly with ethical, social and environmental responsibility. Recent developments at the international level, for example, the Earth Summits in Rio de Janeiro (1992 and 2012) and the Nagoya meeting in 2010 on access to genetic resources and benefits, have sensitized the issue and seen some concrete actions (see Box 2.3). Much more remains to be done, however, in making sure that the genetic resources accounting for the molecular wilderness are fairly and sustainably used.

Molecular genetics and allied biosciences have a large role to play in these efforts. For example, it can help to trace the origins of materials taken from nature and help to determine ownership, or the extent of partial ownership (e.g., genes from plants), of the materials. It can help in creating and maintaining seed banks, tissue culture banks, botanic gardens and other repositories and creating ground rules for material and benefit sharing. Once the active molecules extracted from nature are known and their production as drugs is required, alternative ways of producing them can be made by chemical synthesis or genetic technologies. Such creative contributions to large issues with ethical, social and environmental implications should be the primary concerns of science.

Chapter 3

Drug Targets from Molecular Wilderness

Talent is like the marksman who hits a target which others cannot reach; genius is like the marksman who hits a target, as far as which others cannot even see.
—Arthur Schopenauer, *The World as Will and Representation*, 1819
(translated from German by E. F. J. Payne)

Summary

Infectious diseases represent the dark side of the wilderness which humankind still has to contend with, especially diseases which affect the majority of the world population still living in poverty. The disease pathogens have essential components or processes which can be specific targets for drugs and vaccines. A variety of approaches can be used to identify such targets, including gene knockouts and chemical genomics. Random screening of compounds already available in large pharmaceutical collections can also provide new leads, even when the targets have not yet been identified.

3.1   The Dark Side of the Wilderness

While the wilderness gives us tools to fight ailments in the form of natural products and an understanding of rules of biology, it is

Tapping Molecular Wilderness: Drugs from Chemistry–Biology–Biodiversity Interface
Yongyuth Yuthavong
Copyright © 2016 Pan Stanford Publishing Pte. Ltd.
ISBN 978-981-4613-59-0 (Hardcover), 978-981-4316-60-6 (eBook)
www.panstanford.com

also the source of many ailments, infectious and otherwise. We live in an environment full of other living organisms. As discussed in Chapter 1, many organisms do not cause any harm and may even yield benefits, such as the microbes which live on our skin or in our guts, which do a good job of defending against intruders and yield micronutrients and provide other benefits for us. However, many other organisms bring diseases or disturbance of our normal well-being. To us humans, they are from the dark side of the wilderness.

The general notion that microbes can be the agents of infectious diseases was developed mainly in the 19th century, with crucial discoveries by such notable scientists as Louis Pasteur and Robert Koch. Prior to the general acceptance of what is now known as the germ theory of disease, that is, that infectious diseases are caused by microbial pathogens, the origin of diseases was believed by those in the West to be due to a 'miasma,' or pollution from foul air, contaminated water or rotting organic matter. The word 'malaria', for example, comes from the Italian *mala aria*, which means 'bad air'. These alleged causes were later shown to be just conditions for the spread of diseases brought on by microbial pathogens. In the East, diseases were ascribed to a loss of balance in body and mind, or in 'hot' and 'cold' essences of life (see Chapter 2).

Some infectious diseases come from our direct contact with the pathogens, through air, water, food or human-to-human transmission. The influenza viruses and other microbes which cause respiratory diseases are often spread through air. Tuberculosis can spread when sick people cough tiny droplets containing the tubercle bacteria into the air and people breathe them in. Viruses, bacteria and protozoa which cause diarrhoea often come from food and water contamination. The virus causing AIDS is transmitted mostly through sexual contact. All parts of the body are potential routes for invasion of infectious species. A break in the skin, for example, can allow entry of *Leptospira* bacteria from water contaminated with animal urine, causing leptospirosis, with symptoms ranging from fever and headache to bleeding from the lungs, meningitis and kidney failure. Some protozoa, such as *Negleria*, can cause rare but fatal brain infection by entry through the nose when we swim in the water in which they live.

Once the infectious agents gain access to our bodies, and in many cases to their cellular targets in our bodies, they disrupt our body's functions, often from disrupting the functions of the cellular targets for their own benefits. Viruses and some bacteria gain entry into cells through binding with molecular components of the cell surface, which act as their receptors. Other bacteria secrete toxins which use host molecules as receptors to gain cellular access. Our body responds to these invasions chiefly through immune mechanisms, which include killing the intruders by immune cells, deactivating the molecular components by antibodies or using a combination of mechanisms. For example, some white blood cells can ingest pathogens and kill them by oxidative damage. The defence mechanisms are quite effective in most cases, except when our immune response is weakened by a disease such as AIDS and cannot respond to invasion even by fungi and other agents which are normally unable to penetrate our protective systems.

Some diseases come from transmission by carriers, such as insects and animals. Bubonic plague was a major pandemic on many occasions in history, and since the cause was not then known, it was sometimes ascribed to sorcery, witchcraft or punishment by the Supreme Being. The disease is now known to be caused by a bacterium transmitted by the bite of the fleas which normally infect rats, and past pandemics were likely due to the fact that they turned to humans when too many infected rats, the natural hosts, died. It was not until the late 19th century that Alexandre Yersin [33] and, independently, Shibasaburo Kitasato [34] isolated the causative bacterium, paving the way for full elucidation of the transmission process.

Bubonic plague is now well under control and is no longer a major threat to humans. Not so many other infectious diseases, especially those in tropical areas (see Fig. 3.1). Malaria is a major disease which has also played an important historical role. Alexander the Great and the poet Dante are believed to have died of malaria. In the 20th century, it caused more casualties than direct combat in malarious regions in every war fought. It is now still a major killer, causing more than 500,000 deaths each year, mostly to young children in Africa. The causative agent of the disease, a protozoon of the genus *Plasmodium*, was discovered by Charles Laveran, and its transmission by the *Anopheles* mosquito was shown by Sir Ronald Ross over a century ago [35, 36]. About two weeks after an infective mosquito bite, the malaria parasites, which first develop in the liver, emerge into the bloodstream and invade the red blood cells, causing chills, fever and often cerebral complications. When another mosquito bites the infected person, it takes in the parasites, which undergo sexual multiplication and produce new parasites ready to infect another person.

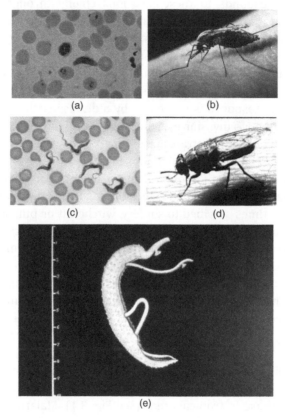

(a)

(b)

(c)

(d)

(e)

Figure 3.1 Some disease pathogens and their insect carriers. (a) Malaria parasites in human blood cells, (b) the *Anopheles* mosquito transmitting parasites during a blood meal, (c) *Trypanosoma brucei*, the agent causing sleeping sickness, swimming in blood, (d) the tsetse fly, the carrier of *Trypanosoma*, and (e) an adult male schistosome carrying a female worm inside its body groove. Pictures from TDR Gallery, WHO.

Sleeping sickness, or African trypanosomiasis, is a fatal disease caused by a protozoon and transmitted by tsetse flies and threatens 60 million people in 36 countries in Africa [37]. The protozoon can pass the blood–brain barrier and cause neurological symptoms, including confusion and disruption of the sleeping cycle, leading to daytime slumber and nighttime sleeplessness and finally coma and death. Another disease, American trypanosomiasis, or Chagas disease, affecting people in Central and South America, is caused by a related protozoon transmitted by another insect called the kissing bug. Left untreated, it can cause life-threatening heart and digestive system disorders.

A large number of people in Africa, South America and Asia are infected by schistosomes, helminthic parasites which enter the human host through skin penetration in water when people take a bath or wash clothes in contaminated water. The parasites migrate to the liver, urinary tract or intestines, depending on the species, to feed on host, mature, reproduce, and lay eggs. Most of the eggs are trapped in the tissues of the infected person, causing inflammation and other complications, ranging from fever to bloody urine to liver cirrhosis. Some eggs released from the host find their way into freshwater snails. Young larvae soon emerge from the snails, ready to infect new hosts and renew the cycle of infection [38].

Many infectious diseases, such as malaria, sleeping sickness, Chagas disease and schistosomiasis, are prevalent in tropical countries, where the majority of people are poor and public health services are weak. The diseases are often complex, requiring not only individual medical care but also public health approaches targeting the whole communities and the environment in order to cut the cycles of infection. Sanitation standards, nutrition and proper health care all play a role in reducing the impact of these diseases. These difficulties are compounded by the fact that they predominantly affect people with little means to secure protection or cures. Because affected people are usually poor, pharmaceutical companies mostly do not put up efforts to develop new drugs or vaccines for these diseases, preferring instead to concentrate on products which bring higher profit margins. These diseases are therefore sometimes called diseases of poverty and—as a consequence—neglected diseases. They call for the attention of the world community at large, not only because of humanitarian reasons, but also because they pose a threat to humankind in general due to greater contacts among people and between people and the environment in the modern world. In short, the world neglects the neglected diseases at its own peril. The problems will be revisited in Chapter 6.

It should be noted that some diseases prevalent in low-income countries are major problems in high-income countries as well, including tuberculosis and AIDS. Emerging diseases like severe acute respiratory syndrome (SARS) and bird flu threaten the whole world community, both due to natural migration of animal reservoirs and due to ease of modern travel. Tropical infectious diseases are appearing more often in the temperate areas because of the increase in travel. On the other hand, diseases which used to be prevalent only in high-income countries, including heart diseases, hypertension, various chronic diseases and many forms of cancer, are now becoming more burdensome in low-income countries as well, partly because people tend to live longer. The health problems of the whole world are therefore interconnected and global effort is necessary to deal with them.

3.2 Fighting the Invaders

Before the nature of infectious diseases and how they spread were known from scientific investigations, we fought against them through knowledge handed down from generation to generation. Traditional medicine gave instructions and formulation of remedies, some of which were similar and others different across various civilizations. Many sought refuge through magic, witchcraft and sorcery. The rise of medical science, together with improved living conditions, brought success to the control of infectious diseases in many countries. At the global level, smallpox, which used to be a disease of major historical importance, was eradicated at the end of the last century through vaccination. Poliomyelitis is also on course to be eradicated through vaccination, although logistic and political difficulties present obstacles in some regions. Vaccines are now available for prevention of some 20 important infectious diseases, ranging from hepatitis B to measles to pneumococcal pneumonia. Even without vaccines, a disease such as guinea worm disease prevalent in east Africa is well on the way to being eradicated, mainly through active case reporting and treatment, access to safe water and community awareness campaigns.

In spite of successes in containing some infectious diseases, however, our fight with many others is hardly over. There are still no effective vaccines against AIDS, malaria and dengue haemorrhagic fever, although experimental vaccines have shown some limited success, prompting the hope for more effective ones in the future. As for tuberculosis, although the BCG vaccine has been long available, it only confers inconsistent protection and better vaccines are clearly needed. There is no effective vaccine against many other infectious diseases, especially the 'diseases of poverty' prevalent in the less developed countries, including sleeping sickness, Chagas disease and schistosomiasis.

Antibiotics have greatly changed the fight against infectious diseases in our favour. As seen in Chapter 2, many natural compounds have the ability to kill microbial pathogens, partly as a consequence of interactions among organisms in the biodiverse environment. Furthermore, in addition to antibiotics from nature, we now have a large number of synthetic drugs, some of which are modelled from natural compounds, while others are totally man-made. They have the capacity to kill pathogens and cure many infectious diseases, ranging from diarrhoea to respiratory and other infections. Antibiotics are mostly effective against bacterial infections, but there are drugs against fungal and parasitic diseases as well as some which are effective against some viral infections. There is, however, a serious flaw in the use of antibiotics, which demonstrates the untamed nature of the wilderness. Up to now, soon after any antibiotic was introduced, resistance arose, necessitating a search for agents to control the resistance or for completely new drugs altogether. The ability of the wilderness to fight back our efforts to 'tame' it leads to the concept of not always using brute force of technology but also striking a balance with nature and an ecological approach to health and diseases.

In general, vaccines against infectious diseases act by boosting our immune system to fight against pathogens, while drugs act by killing or debilitating pathogens so that they no longer exist in our body, or at least do not pose a threat to our health. Other drugs called prophylactics act to prevent new infection, and yet others act to relieve the symptoms of the diseases. Many drugs and vaccines act specifically on the essential components of disease pathogens. We call these components the targets. While many vaccines are directed at enhancing immunity against pathogens in general, for example, by using a weakened pathogen which the body can get rid of and learn to mount a protective effect against the real invading pathogen, other vaccines are aimed specifically at enhancing immunity against particular components of a pathogen, that is, they are target-directed. Drugs can likewise act at many points to interfere with the disease process, but target-directed drugs will achieve the specific purpose of curing the disease by stopping the pathogen at specific points through inactivating specific molecules essential for its survival. Borrowing from Greek mythology, a target-directed drug is an arrow directed at the Achilles' heel of the pathogen. But perhaps we can turn to a story from the East to understand drug targets better.

3.3   The Concept of Drug Targets

The Ramayana epic tells the story of combat between Rama, the avatar of the god Narayana, and Ravana, the demon king who abducted Rama's wife. Rama could not kill Ravana no matter how hard he tried, until he learnt that Ravana had deposited his heart with a holy man, thereby making him impossible to kill. Only after Hanuman, Rama's monkey general, succeeded in obtaining the heart of Ravana, could Rama succeed in killing his enemy.

We might compare this story to our own quest in defeating a disease pathogen. Until we can find the 'heart' of the pathogen, we will not be able to defeat it. In short, the 'heart' of a disease pathogen is its underlying essential component or process without which it cannot survive. We can aim to kill it by destroying such a component or process with a drug or some other agent of interference. This comparison, however, is a simplification. In truth, a pathogen has many essential components or processes, not all of which may be accessible or vulnerable to disruption by drugs or other forms of interference. Moreover, there may be similar components or processes within us, the host bearing the pathogens, which are also similarly vulnerable to such disruption, hence causing toxic or undesirable side effects of the drugs or the interference.

The concept of hitting specific targets of the pathogens with drugs or drug combinations, thereby achieving the goal of curing the disease and regaining health for the host, carries two main aims. One is to have agents with high efficacy, which act at the targets effectively, consequently resulting in the death of the pathogens. Another is to avoid the toxic or side effects of the drugs on the patient, that is, to have selectivity of the drugs for the pathogens only. This concept was first developed over a century ago by Paul Ehrlich, who said that every disease should be treated with a chemical specific for that disease [39]. Ehrlich came to this conclusion after he discovered the antimalarial effect of methylene blue, which could stain the malaria parasite, indicating its potential affinity for the parasite target. He later discovered the arsenic-containing drug salvarsan as an effective agent against syphilis, with far fewer side effects than mercury salts, which were used for conventional therapy at that time. Considered the father of chemotherapy, Ehrlich reasoned that a specific cure for an infectious disease can be found if a toxic agent can be preferentially directed to the pathogen with little effect on the host. He called this ideal therapeutic agent the 'magic bullet'. In his own words [39]:

> In order to pursue chemotherapy successfully we must look for substances which possess a high affinity and high lethal potency in relation to the parasites, but have a low toxicity in relation to the body, so that it becomes possible to kill the parasites without damaging the body to any great extent. We want to hit the parasites as selectively as possible. In other words, we must learn to aim and to aim in a chemical sense. The way to do this is to synthesize by chemical means as many derivatives as possible of relevant substances.

This concept has been explained today in terms of specificity of interaction between the therapeutic agent and the molecular target in the pathogen. The lack of toxic effects on the host is explained by the absence of the corresponding target in the host, or more often differences between the corresponding molecules in the host from the pathogen. In the latter case, only the target molecule in the pathogen is hit and disabled by the therapeutic agent, while the molecule in the host, which may be functionally or structurally similar, is different enough not to be affected, or affected only mildly.

3.4   Targets and Receptors as Crucial Components of Life Processes

We have come a long way from the days when Ehrlich conceived of the idea of specific targets for drugs. Thanks to advances in biochemistry and molecular biology, we now know the structures and functions of various parts of most important pathogens in detail, down to the molecular level. Although we do not know the molecular targets of all drugs, since many drugs act at different sites simultaneously, numerous targets of drugs have been identified and studied in detail. The functions and modes of action of many targets are also known. In many cases we can use the knowledge about drug targets to design new drugs and to understand the mechanisms of drug resistance when it occurs.

Drug targets are generally defined as molecules or molecular complexes, the functions of which are modified by drugs through specific interactions. Generally a drug specifically binds a target and modifies its function by inhibiting or stimulating it or changing its activity in some other ways so as to result in an effect on the organism carrying the target. If the organism is a pathogen, it may be killed or debilitated. In other cases, such as in noninfectious illnesses, the drugs are directed to the targets within the patient, resulting in desirable changes and therapy.

Drug targets can be classified into several types [40]:

- Enzymes (proteins with catalytic functions), normally acted upon by small-molecule drugs.
- Substrates, metabolites and other proteins, the first two of which may be small molecules, and the drugs may be enzymes.
- Receptors, such as hormone receptors. A family of receptors called the G-protein-coupled receptors are targets of 30–40% of all known drugs. Many of these receptors have important roles in the central nervous system (CNS).
- Ion channels and transport proteins, normally located on membranes.
- DNA, RNA and the ribosomes, involved in gene functions and production of proteins.
- Protein–protein and protein–nucleic acid interacting interfaces.
- Complexes of proteins, and proteins with DNA, RNA and other biomolecules.
- Other types of molecules or molecular interactions, such as those which are involved in the structure of membranes and other organelles.

An ideal drug should interact and modify only the targets, which leads to intended effects only and not to side effects due to interaction with unintended targets. For therapy of infectious diseases, a target for which drugs are aimed, such as an enzyme in a biochemical pathway or a structural component of the pathogen, should ideally not also be present or should not be an essential component in the host as well. Alternatively, if an analogous enzyme is present in the host, it should not be susceptible to inhibition by the drug to the same extent, or should be only inhibited to a minor, insignificant extent, as otherwise toxicity or undesirable side effects can be anticipated. In this respect, the words of Paracelsus uttered five centuries ago should be heeded: 'All substances are poison; there is none which is not a poison. The right dose differentiates a poison from a remedy.' Luckily, today we have many drugs which are not poisons even at relatively high doses, but we still need to be careful in developing or using them.

How does a drug specifically affect a target, resulting in therapeutic effect? It can do so by a variety of mechanisms, some of which are shown diagrammatically in Fig. 3.2. In this case the drug is an inhibitor of its target enzyme. The target enzyme normally functions in producing, say, metabolites essential for the life process of a pathogen through catalysis of the reaction of the precursor

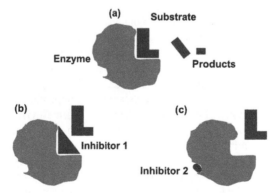

Figure 3.2   A drug may act as an enzyme inhibitor, inhibiting its target enzyme and preventing its normal function in catalysing reactions of its substrates, hence preventing product formation. (a) Normal function of the enzyme; (b) an inhibitor, 1, binding at the active site, preventing substrate access; and (c) another inhibitor, 2, binding at another site, causing the active site to change and unable to accommodate the substrate.

substrate. An inhibitor may act as a drug by specifically binding with the active site, blocking the substrate from access to the site. Another inhibitor may not bind at the active site but at another site which causes a change in the enzyme structure so as to be unable to accommodate the substrate. An example of a drug binding at the active site of the target enzyme, preventing its normal function, is shown in Box 3.1. The target enzyme in this case is a malarial enzyme called dihydrofolate reductase, involved in converting dihydrofolate to tetrahydrofolate, for subsequent use in making such metabolites as thymidylate. The inhibitor is pyrimethamine, which exerts an inhibitory action on the enzyme from malaria parasites but not on human enzymes, therefore explaining its specificity.

Box 3.1   An Enzyme Inhibitor as a Drug

Working on nucleic acid metabolism, George Hitchings [48] and his colleagues at the pharmaceutical company Burroughs Wellcome developed a system to test the antibacterial acitivity of various compounds. Among these, compounds in the 2,4-aminopyrimidine family were found to inhibit *Streptococcus faecium* effectively, but the inhibition was prevented when folinic acid (a tetrahydrofolate derivative) was present. From this and other lines of evidence, they concluded that these compounds acted through inhibition of the enzyme dihydrofolate reductase, which supplies tetrahydrofolate to the organism. They extended the investigation to malaria and found that compounds such as pyrimethamine, a 2,4-diaminopyrimidine, have good antimalarial activity by virtue of their inhibition of this enzyme.

Dihydrofolate reductase plays an essential role in producing tetrahydrofolate required for synthesis of deoxythymidine, an important component of nucleic acids, as well as for production of other essential metabolites involving transfer of 1-C (e.g., methyl) groups. It is an essential enzyme in a wide variety of organisms, including humans. How, then, can it be a drug target against infective microorganisms? How can drugs inhibit it and exert antimicrobial effects without inhibiting the host enzyme and producing toxic side effects? The answer lies in the fact that although enzymes from various organisms share similar basic structures, there are large differences both in the amino acid sequences and the three-dimensional structures, both as a whole and of the active sites in particular. Inhibitors can therefore be found which inhibit only the microbial enzyme without inhibiting the human counterpart.

(Continued)

Box 3.1 (*Continued*)

Figure 3.3 shows the structure of the active site of malarial dihydrofolate reductase bound with its substrate, dihydrofolate, and with its inhibitor, pyrimethamine, as determined by the X-ray diffraction technique [49]. Pyrimethamine binds the active site avidly, thereby preventing binding of the substrate and inhibiting the action of the enzyme.

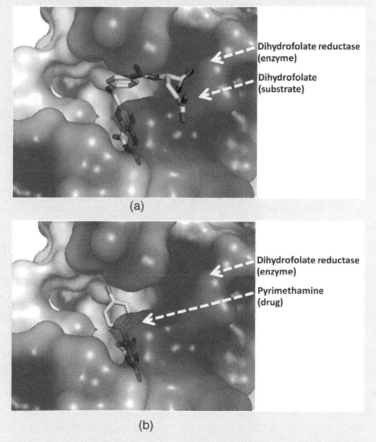

Figure 3.3 (a) The active site of malarial (*Plasmodium falciparum*) dihydrofolate reductase is a cleft which accommodates its substrate dihydrofolate. (b) Pyrimethamine binds in the same place as the substrate in the active site, thereby preventing the binding of the substrate and the normal reaction. Pictures courtesy of Dr Penchit Chitnumsub, BIOTEC Thailand.

3.5 Finding the Right Targets: Classical and Chemical Genetics

It may seem straightforward in drug development to identify specific targets and find compounds which would preferentially hit or inhibit the targets without affecting the host. In reality this is not so easy. The targets have to be identified and shown to be susceptible to the drugs directed against them. At the same time, the drugs must not have side effects on the host targets. In the case of pathogens, it may be relatively straightforward to identify the targets which are not present in the host or present but with different susceptibilities to the drugs. In diseases which result from abnormality of the host, the target may be a molecule, say, a hormone receptor with an abnormal function. In such cases, drugs have to be found which only affect the abnormal component without side effects on the normal one.

Advances in genomic technologies have enormously helped in the effort to identify drug targets. We now have information on various genes of a living species. In the classical genetic approach, genes are studied by, say, mutation which produces effects on their functions. In many cases, the suspected target gene might be 'knocked out' (e.g., disabled or deleted) and the effect examined. Knocking out an important target, for example, an enzyme in a pathogen which is important in producing some essential metabolite for the pathogen, may make it nonviable, and it may have to be rescued by exogenous addition of the metabolite. More recent methods of identifying essential targets include modification of gene expression or temporarily shutting off a gene's activity and observing the effects of loss of gene function. These are some of the means by which drug targets can be identified.

With the current availability of proteomic information from cells bearing diseases and many infectious organisms, drug targets may be identified by a process of reverse pharmacology. In an approach called activity-based proteomics or activity-based protein profiling [41, 42], a compound which is known to interact with proteins which may include desired targets is chemically bound to a tag which can be identified by fluorescence or other convenient means and used as a probe. It is applied to cells or organisms of interest, and all the proteins which react with the probe can be identified. The profile of tagged proteins will be changed with the action of drugs or compounds, selectively inhibiting one or more of the proteins, which can then be identified as potential targets.

In what is called the chemical genetic approach, small molecules, collections of which are called probe libraries, are added to cells or purified proteins, and the effect is studied by biological assays in test tubes or animals. Once a drug-like effect is established, the target protein(s) and gene(s) can be identified by methods such as activity-based proteomics, or activity-based protein profiling, which can identify the proteins which are possible drug targets through the use of probes. To identify the targets more precisely, more compounds can be tested, either from existing chemical libraries or from newly synthesized ones, on both the total protein complement or on cells. The iterative method eventually leads to drug candidates to be selected for further studies. As an example, the pain reliever aspirin (acetylsalicylic acid) was found to inhibit the action of cyclooxygenase (COX) enzymes, which catalyse the synthesis of prostaglandins, the pain signals. This led to the identification of the genes for COXs as targets for pain-relieving drugs [43].

Drug targets may also be identified by pinpointing DNA changes responsible for drug resistance. As will be seen in Chapter 5, drug resistance often occurs from mutations in the gene of the target so as to reduce the effect of the drug, or amplification of the target gene so as to swamp the effect of the drug. Comparison between DNA profiles of sensitive and resistant cells, either through microarrays or though whole-genome sequencing, can pinpoint the candidate genes with changes which give rise to drug resistance, and which may therefore be the drug target genes. This approach has been used to identify the target gene of the antimalarials spiroindolones as an ATPase enzyme important in cation transport [44].

A combination of two approaches is complementary and paves the way for identification and verification of the molecular targets. At the risk of oversimplification, we might say that in classical genetics we first use mutational and other genetic studies to identify drug targets, from which small molecules will be developed with specific action against the targets. Chemical genetics, on the other hand, uses chemicals to exert specific actions and hence identify the targets, against which more molecules can be developed with higher specificity and efficacy. Furthermore, gene families from which the target gene was identified can also be investigated as possible targets for an understanding of their functions, the effect of the drugs and development of other drugs. Figure 3.4 gives a summary of the two complementary approaches to target finding and drug discovery.

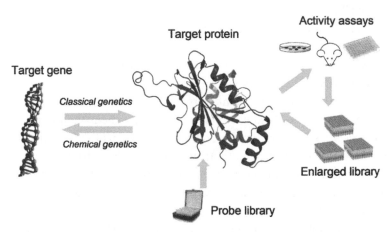

Figure 3.4   A simplified diagram showing classical and chemical genetics approaches to finding of a drug target, and development of chemical libraries, from which drug candidates can be selected. In classical genetics, the suspected gene is modified and the effect on the suspected target protein studied by suitable assays. In chemical genomics, a probe library of chemicals is used to study the effect on the suspected target protein. The corresponding gene can be identified, and the biological effect of the chemical probes can be studied and analysed, leading to design and synthesis of larger libraries, from which suitable drug candidates may be identified.

It must be pointed out that identification of drug targets is only the beginning of the long search for effective compounds which hit the targets. The compounds must be bioavailable and reasonably stable, namely, they must be able to access the targets and interact with them long enough to exert the intended biological effects. They must also not have side or toxic effects on the host, nor should their metabolites have such effects. In short, the targets must be 'druggable', that is, susceptible to development of suitable drugs, which must also have drug-like properties.

3.6 Natural Products as Underexplored Sources of Drugs and Tools for Drug Target–Finding

Currently, except for only a few herbal drugs in the pharmacopoeia of developing countries, natural products form only a small fraction of drugs in current use. There have, however, been some spectacular successes in developing natural compounds as drugs. The family of lactam antibiotics, including penicillin discovered by Alexander Fleming, was found to act by inhibiting the bacterial enzyme transpeptidase, which is used in making the bacterial cell wall. More recent examples include the antibiotic glycopeptide vancomycin, which also inhibits cell wall synthesis, and Taxol, which targets the cell's microtubules. Compared with drugs developed from chemical synthesis, however, natural products are still in the minority. They tend to have complicated structures and are difficult to synthesize. Many of them have structures which dictate against their drug-like properties.

Nevertheless, a main advantage of natural products is that they can serve as template molecules, from which more drug-like compounds can be synthesized. In the case of penicillin, for example, new synthetic or semisynthetic drugs (synthesized from building blocks provided by fermentation) have been made based on the parent structure, which have superior activities and are effective against penicillin-resistant strains of bacteria. Even without knowledge of the target of action, structurally related drugs can be studied, and structure–activity relationships found from such an undertaking can lead to a better drug.

Apart from optimization through chemical modification, natural products are also often used as the starting points for discovering new drugs through drug target identification by reverse pharmacology and chemical genetics, as described in the preceding section.

We have already seen from the preceding chapters that natural products are parts of the molecular wilderness in nature, which are used by living organisms to interact with other organisms, in both hostile and beneficial ways. Other natural products act as messengers or control switches for the organism's own function. As relatively small molecules, they often work by binding to specific target molecules, often a protein or a group of proteins located in the responsive organs or tissues. An example is the binding of nicotine, a natural product from the nightshade family of plants and also in tobacco, to receptor molecules in the nervous and muscular systems. The interaction leads to increases in the level of some neurotransmitters, which are responsible for the accompanying sense of euphoria and relaxation. Another example is the identification of opioid receptors as targets of opiate drugs like morphine. Since they have evolved to interact with other natural molecules, natural products can be therefore considered as nature's tool for hitting molecular targets. We can therefore use them, not only as drugs, but also as tools for finding drug targets.

3.7   Hitting the Targets: Drugs by Design

Once the target or targets for drugs have been identified, a rational design of new drugs is possible. New drugs are needed for many reasons. Many viral diseases are not susceptible to treatment with currently available drugs, and those which are normally adapt themselves so quickly that new drugs are continuously needed to fight them. In general, drugs are not always effective; normally only 30–70% of patients respond positively to any drug because of differences in their genetics and other characteristics. In an extreme case, the drug melarsoprol, an arsenic-containing drug for treatment of second-stage human sleeping sickness, is highly toxic and can cause up to 10% fatalities and is used only because of a lack of better therapies [45]. Drugs such as these should be replaced by new, more effective drugs, preferably designed against specific targets, with less severe side effects. Eflornithine, directed against the parasite ornithine decarboxylase target, is one such drug. Called a 'suicide substrate', it has a structure similar to ornithine, the substrate of this enzyme, but will bind covalently to it and stop its function. Originally developed for cancer treatment and found to be effective against some forms of sleeping sickness, the drug went on market in the nineties for a while but was stopped due to lack of profit. Fortuitously, the drug has another property, slowing of facial hair growth, which made it marketable, and arrangements were eventually made by the drug company to produce it also in lifesaving formulations as donation aid to countries where sleeping sickness is endemic.

Even when effective drugs exist, there is still a need for new drugs in the pipeline. In the case of anti-infective drugs, resistant pathogens will inevitably arise from mutation of the target gene, or other mechanisms. Drugs for noninfectious diseases sometimes also lose their efficacy due to various tolerance mechanisms. The drugs themselves may also have limitations, such as poor absorbance from the digestive tract or a short lifetime in the bloodstream. These obstacles can be overcome by modifying the structures, changing the mode of drug delivery or making other changes.

A typical process for target-directed drug design is given in diagrammatic form in Fig. 3.5. Once the target has been identified, probably from a combination of classical and chemical genomics as outlined above, its three-dimensional structure should be determined by X-ray diffraction or other suitable methods, preferably in bound form, with attached substrates or inhibitors. Even when the target structures cannot be determined directly, it is often possible to derive good approximations through computer-based modelling, making use of known amino acid sequences and known structures of similar biomolecules; however, this approach is more error-prone. If there is information on, say, inhibitory activities of structurally related inhibitors, the quantitative structure–activity relationship can be used to obtain a good binding model from which it is possible to predict effects of modifying various parts of the structure. The main criterion for potentially good inhibitors is that there should be good affinity for binding, measured as free energy of binding, which is composed of two parts, enthalpy and entropy. Often the enthalpy of binding is the major factor in optimization of a potential drug structure [46]. In many cases, drugs can be designed by combining optimized potential fragments into whole molecules.

The aforementioned processes form the basis for design and synthesis of new molecules with better activities than the original ones, and hence an enlarged library of potential drugs can be obtained. The library members are examined for their biological activities against the target, and the promising members can be examined for drug-like properties, such as absorption, distribution, metabolism and excretion (ADME). The data are used to improve design and make a more refined library of potential drugs, which can be assessed for their activities in more detail with, say, an animal model. The interesting molecules are also subjected to exploratory toxicological assessment, including mutagenic and other undesirable side effects. These steps are repeated in iterative cycles, using information obtained from each cycle to refine the molecular structures, until eventually a small number of potential candidates is derived, which can proceed to more detailed preclinical testing. The molecule with the best drug properties may be selected for clinical testing and further development.

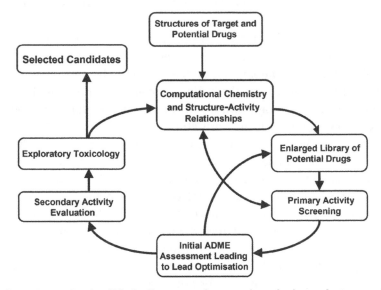

Figure 3.5   A simplified diagram of target-based drug design and development. The structures of the target, such as an essential enzyme of the pathogen, and potential drugs directed against the target, such as an inhibitor of the enzyme, obtained by such techniques as X-ray diffraction are analysed through computational chemistry in conjunction with information on structure–activity relationships of various inhibitors. Iterative cycles, consisting of design and synthesis of libraries of potential drugs on the basis of the foregoing analyses; activity screening; assessment of drug-like properties, such as absorption, distribution, metabolism and excretion (ADME); and exploratory toxicology lead to selection of a few candidates for further detailed investigation as potential drug candidates.

It is clear that the processes for target-based drug design and development are long, and full of uncertainties. Very often, no suitable molecules can be found. In a few cases, however, the efforts are rewarded by selection and development of new drugs with desired indications. Some examples of target-based designed drugs are given in Box 3.2. It should be noted that even the drugs which have been developed and marketed are still subjected to continuous surveillance in order to make sure that they are free from unwanted effects, and new information may come to light which necessitates their withdrawal in a few cases and improvement in many others.

Box 3.2 Examples of Drugs from Target-Based Designs

The influenza virus uses its enzyme neuraminidase to release it from the host cell in the budding process after reproduction in its host. The enzyme cuts the link between the new budding virus particles made in the host cell and the host cell membrane through the host's sialic acid, which the virus binds with its receptor so that the virus particles can go free (Fig. 3.6a). Drugs such as Relenza (zanamivir) and Tamiflu (oseltamivir) have been designed as anti-infuenza agents by virtue of their ability to interfere with the action of this enzyme [50, 51]. Both drugs have similar structures to the substrate, sialic acid. Relenza and the active form of Tamiflu inhibit the enzyme by binding to the active site in a similar manner, with consequent antiviral action. Tamiflu is a pro-drug, which has to be activated first in the body through cleavage of an ester bond, thereby releasing the active drug. By inhibiting the enzyme, the virus particles are stuck and the infection is stopped. However, the virus can fight back by developing resistance to the drugs through mutation at the active site, which weakens the binding of the inhibitors.

Aspirin (acetylsalicylic acid, Fig. 3.6b), a natural product found in willow bark, a drug long used for relief of pain and inflammation and also against blood clotting, acts mainly by inhibiting the enzyme COXs involved in the production of prostaglandins and thromboxanes responsible for the inflammation and clotting processes, respectively. There are two main forms of COX (COX1 and COX2), both of which aspirin inhibits indiscriminately, accounting for many side effects. Drugs specific for inhibition of COX2, mainly accounting for pain relief, should have less severe side effects. Such drugs have been designed and made, some of which specifically inhibit COX2 by binding in the COX2 active site pocket which can accommodate them well; however, they do not bind to COX1 due to the presence of bulky side chains,

(Continued)

Box 3.2 (*Continued*)

which cannot be accommodated by COX1 [52, 53]. However, some of these new drugs have been shown to have other unexpected side effects, which necessitates a further search for even better drugs. These examples illustrate difficulties in finding new safe drugs, including target-directed ones.

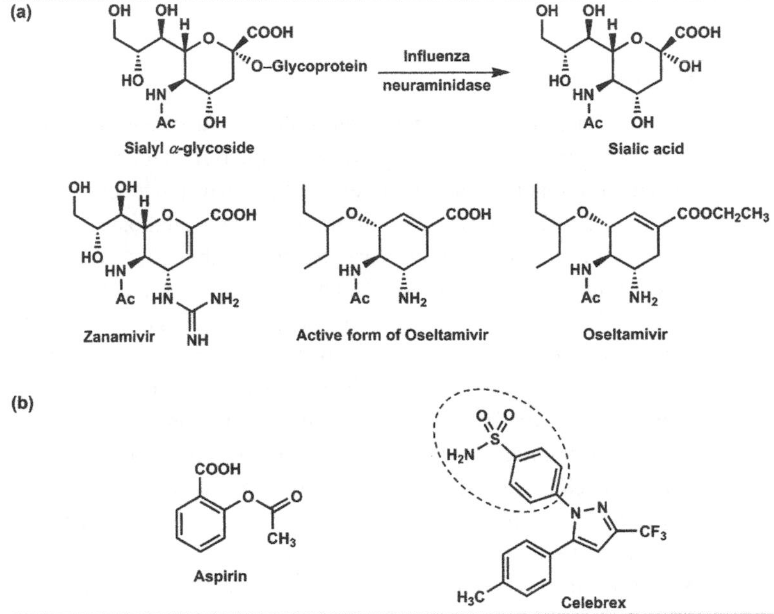

Figure 3.6 (a) Influenza neuraminidase is important for virus reproduction through its action in freeing the virus from its host receptor by hydrolytic release of sialic acid. Zanamivir and oseltamivir act by inhibiting this process. (b) The structures of pain relievers aspirin, which inhibits both COX1 and COX2, and Celebrex, which inhibits only COX2 specifically due to the presence of a bulky side chain (in dashed circle) which cannot be accommodated by the COX1 active site.

In addition to targets of drug action, biomolecules can also be used as targets for drug delivery. In targeted drug delivery, antibodies raised against, say, membrane receptors of the cells where drugs are to be directed, are used as homing devices for drugs incorporated together in small packages, such as liposomes, micelles or nanoparticles. Here the targets are markers for delivery of conventional drugs preferably to diseased cells. This approach helps to reduce unintended damage to healthy cells and thus reduces toxic effects from mistargeted drug distribution.

3.8   Hitting the Targets: Random Screening

In contrast to rational design, drugs can be discovered by random screening of compounds against infectious microbes or animals with appropriate disease models. The discovery process is much aided, however, when the molecular targets are known. Random screening of chemical libraries, or collections of chemicals, is a powerful tool in drug discovery. The efficiency of the process is increased by introduction of high-throughput assay methods so as to enable the screening of tens of thousands of molecules in the collection of pharmaceutical companies or research laboratories in reasonable periods of time. The 'chemical space', encompassing total possible structures of compounds with molecular weights less than 500, is very large and has been estimated to contain between 10^{23} and 10^{60} molecules [47]. There is therefore a possibility that suitable drugs exist among this vast number of possible structures, if only they can be found. However, to find a good 'needle in the haystack', preliminary work should be done to increase the possibility. In some cases, the binding to the intended target of virtual libraries of small molecules, from computer-generated structures, can be calculated first so as to have estimates of the importance of various groups of various compound types. Virtual screening of compounds, done through fitting of the computer-generated virtual compounds with the active sites of the biological targets, is a useful and relatively inexpensive method for gaining early information on potential lead compounds, which can then be actually synthesized and tested.

3.9   Phenotypic versus Target-Based Screening

Random screening is a useful tool in drug discovery, even when the molecular targets against which the drugs are directed are not known. As seen in Chapter 2, natural compounds can be directly screened against disease pathogens or against diseased cells, tissues or whole animals. This is called phenotypic screening, as distinct from target-based screening. Such screening can be done in a high-throughput format, such as a cell-based assay with multiwell plates. Indeed, many drugs were discovered by this trial-and-error method, including many antibiotics and anticancer drugs from microorganisms and other sources.

Phenotypic and target-based screening approaches are often done in conjunction so that hits from target-based screening also show desired biological activity in phenotypic screening. Conversely, active compounds from phenotypic screening need to be examined for their mechanisms of action, including the nature of the targets on which they act so that they can lead to optimized compounds with high efficacy and low toxicity. They can also be done in succession so that only a few promising compounds from one screening are subject to the other.

The compounds found to be active in these screenings are only hits, which need to be optimized. There are ways to generate improved compounds, which can become leads to new drugs. These leads must undergo further optimization for efficacy and specificity of action, after which preclinical and clinical drug development can begin.

Chapter 4

Molecular Wilderness as Templates for Drugs

You look at where you're going and where you are and it never makes much sense, but then you look back at where you've been and a pattern seems to emerge. And if you project forward from that pattern, then sometimes you can come up with something.

—Robert M. Pirsig, *Zen and the Art of Motorcycle Maintenance: An Inquiry into Values*, 1974

Summary

Original molecules from nature provide templates which can be scaled up or modified to make better drugs through chemistry, biology and allied sciences. Molecular diversity from combinatorial and diversity-oriented chemical synthesis provides even wider selections. Fragment-based drug discovery shows the power of making effective drugs from components, each of which may bind only weakly to the target. Combinatorial biosynthesis provides a method for producing 'nonnatural' natural products, while metagenomics can lead to the discovery of new antibiotics, even from microorganisms which cannot be grown. Rules governing the ability of drug molecules to act effectively, including surviving long enough in the host and accessing tissues and targets of action, can be

Tapping Molecular Wilderness: Drugs from Chemistry–Biology–Biodiversity Interface
Yongyuth Yuthavong
Copyright © 2016 Pan Stanford Publishing Pte. Ltd.
ISBN 978-981-4613-59-0 (Hardcover), 978-981-4316-60-6 (eBook)
www.panstanford.com

used to build better platforms for development of new drugs from original natural molecules.

4.1 Expanding the Potentials of Molecular Wilderness

Although the range of molecular wilderness is vast in giving us drugs against various ailments, there are limitations which need to be overcome in order to realize its full potential. Take the case of antibiotics, natural products produced by microbes, plants and other organisms against other microbes which are pathogens. At least three main problems were encountered from the beginning of their discovery.

First, the quantities of antibiotics produced in nature are small and normally insufficient to meet the demand for human therapeutic use, especially in their extraction and purification from natural sources. Industrial production of these antibiotics needs to be devised in order to enable them to have widespread use. The best microbial or other sources of antibiotics need to be found. Fermentation technologies for the production of drugs also need to be developed. The history of the production of penicillin provides a lesson of how these problems were overcome [54, 55]. After the chemical nature of the compound was established and definitive proof of its value in treating infections was obtained, effort was undertaken to find the best source of the *Penicillium* mould and to develop the fermentation process which would yield large amounts of the drug. This was accomplished just in time for the large demand as the Second World War reached a peak before its end. Penicillin made a major difference in reducing wound infections and therefore the number of deaths and amputations resulting from the war. Today, industrial production of antibiotics in penicillin and other families by fermentation makes use of optimized strains of producing microorganisms, often with gene replacements for more effective production, and the processes are highly automated for control of various contributing factors [56].

Many drugs from plants have to be harvested from their natural sources, which can be tedious and expensive. Some plants or parts thereof, from which the products are derived, have to be supplied from their natural sources, posing threats to environmental sustainability. Others which can be harvested from suitable plantations can impose a high cost on production. Paclitaxel (Taxol) is an anticancer drug derived from endophytic fungi in the bark of the Pacific yew tree. However, it has been estimated that treating a patient would require the consumption of eight 60-year-old yew trees [57]! Such high environmental and financial costs must be avoided in an attempt to obtain new drugs from nature.

Secondly, although antibiotics from nature, like penicillin, are very effective against a number of infectious organisms, they have a narrow spectrum of activity; namely, they are only effective against a limited range of such organisms. The mechanism of action of penicillin (see Box 4.1) provides a way to expand the spectrum of action through chemical modification of the drug to make semisynthetic penicillins, which have better properties, such as greater accessibility to the microbial pathogens inside the host and an ability to treat a wider variety of infections, including penicillin-resistant infections. The key step to semisynthetic penicillins is the production of 6-aminopenicillanic acid (6-APA), which retains the beta-lactam function necessary for efficacy and is the nucleus for adding other chemical groups which give additional properties to the semisynthetic drugs. Synthetic modification of natural products derived from plants can also result in new drugs with enhanced qualities, such as artesunate, an antimalarial modified from artemisinin extracted from the *Artemisia* plant, with higher solubility and better oral bioavailability.

Thirdly, the chemical diversity of antibiotics needs to be expanded so as to increase the ability to deal with various infectious diseases. The pioneering work of Selman Waksman, who coined the term 'antibiotics', led to the discovery of streptomycin from *Streptomyces* bacteria [27]. General methods for screening microorganisms for the presence of useful antibiotics were developed, which led to the discovery of many more antibiotics with various modes of action. At present, microorganisms from soil and various sources in the environment have been screened exhaustively by various groups of researchers and drug companies. New antibiotic discoveries have become increasingly rarer after the first few decades since the 1940s, and so new sources need to be tapped, including the deep oceans and other hard-to-reach locations.

Box 4.1 Making Use of the Microbial Wilderness

Antibiotics are substances which microorganisms produce to kill other microorganisms in their habitats, that is, their competitors or their prey. We have learnt how to tap the potentials of microbial weapons for our own fight against infectious diseases. Starting with the discovery of penicillin, other antibiotics were soon discovered and many were produced as drugs. Figure 4.1 shows the structures of some drugs in the penicillin family. They all have a beta-lactam function (dashed circle) in their structures, which exerts its action by binding and inhibiting penicillin-binding proteins (PBPs) or transpeptidases, the enzymes responsible for joining peptide bonds in the formation of the bacterial cell wall. Different penicillins were made, mainly through semisynthetic processes, starting with 6-aminopenicillanic acid (6-APA), which in turn is made from enzymatic hydrolysis of an extracted penicillin. The different penicillins were needed in order to widen the spectrum of action, increase the bioavailability or modify drug properties in other ways. The need for new antibiotics also stemmed from the fact that the pathogens acquired resistance to the drugs soon after widespread use, a constant feature of wilderness struggles, which will be explored further. Cephalosporins, the parent compound of which was discovered from the fungus *Acrimonium*, and carbapenems, originally from *Streptomyces*, represent newer antibiotics from the beta-lactam family. Other antibiotics with modes of action different from penicillins were discovered from soil bacteria and other sources, including the aminoglycosides streptomycin and gentamycin, which work through inhibition of bacterial protein synthesis.

Mass production of antibiotics is a major task. This required extensive effort and advances in biochemical engineering and fermentation technology. Industrial production of penicillins, for example, was achieved through the use of aerated deep fermentation with corn steep liquor as a growth medium. Although the original definition of an antibiotic is a substance made by a living organism to kill other organisms, it has now been expanded to include semisynthetic and completely synthetic substances with such activities. Some antibiotics, such as quinolones, are made entirely from chemical synthesis.

(Continued)

Box 4.1 (*Continued*)

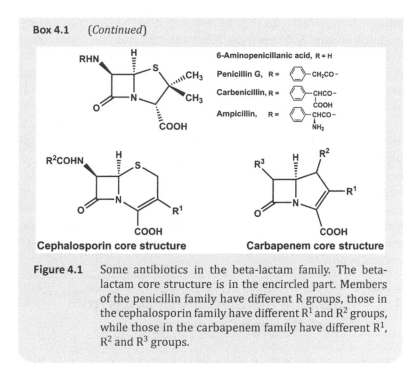

Figure 4.1 Some antibiotics in the beta-lactam family. The beta-lactam core structure is in the encircled part. Members of the penicillin family have different R groups, those in the cephalosporin family have different R^1 and R^2 groups, while those in the carbapenem family have different R^1, R^2 and R^3 groups.

4.2 Sustainable Production of Drugs from Nature

To obtain drugs from nature, the conventional way is to access the sources, extract them and purify them. While sources such as microorganisms can be isolated and produced on a large scale by industrial fermentation, plants and exotic sources such as sea sponges present problems for sustainable production of the desired compounds. An example mentioned before is production of paclitaxel (Taxol), a mitotic inhibitor used in cancer treatment, which originally required harvesting the bark of the Pacific yew tree for extraction and purification of the compound. This resulted in the death of the trees and posed an ecological problem, which was gradually solved through various means [57]. Plantation of *Taxus* trees was undertaken so as to be able to harvest them sustainably. A semisynthetic method was also developed so as to avoid harvesting the bark but harvesting the needles instead. These contain an intermediate, which can then be converted to the final product by chemical means. Subsequently, plant cell cultures were developed to replace harvesting of the products from the trees altogether. The yields were improved by optimization of the process, which was assisted greatly by the discovery that endophytic fungi can produce, or help to produce, Taxol.

Artemisinin is an antimalarial which is produced by *Artemisia annua*, or sweet wormwood. Although the supply can come from plantations, increased need for artemisinin and its derivatives necessitates the search for alternative means of production. This has been provided by using yeast transformed with the genes obtained from the plant to produce artemisinic acid, from which the final product can be made [58]. This is an example of new solutions made possible by new genetic technologies for sustainable production of natural products.

4.3 Expanding the Diversity of Drugs from Nature through Chemistry

The diversity and access to molecules from nature can be expanded with the help of chemistry. The diversity of compound collections, called 'libraries', can be vastly increased through chemical synthesis using the original molecules from nature as templates. As mentioned in Chapter 3, there is a vast number—a large chemical space—of possible compounds which can be explored as potential drugs. Since natural products form a rich collection of compound classes with diverse structures evolving by natural selection to have specific biological activities and functions, they are ideal starting points to explore chemical space with potential for drug discovery [59–61]. Starting from the structure of a natural product, preferably one with already some activity against a disease target, libraries can be made and explored, such as for random screening as an initial step in drug discovery or for optimization of the potential drug candidates. Even when the molecular target of action of a natural product is not known, when its efficacy against a disease is known from traditional medicine, for example, the basis of action can be explored at the cellular or the whole-organism level. These original molecules from nature provide templates, which can be modified to make better drugs. Many natural products active against malaria, for example, have served as starting points for future therapies [62].

Techniques embodied in the process called combinatorial synthesis [63–65] can be used to generate families of structurally related compounds for such purpose. Combinatorial synthesis is the synthesis of a library of a large number of compounds through a stepwise reaction with different building blocks. The field started with the development of techniques for making a large number of peptides attached to solid beads. In building a peptide library, one of 20 common amino acids can be coupled with another amino acid, giving a total of 400 dipeptide products. The next coupling reaction gives a total of 8000 tripeptide products. In general, for n building blocks, the number of products will increase by n times with each round of the coupling reaction. The principles of combinatorial synthesis are explained in more detail in Box 4.2.

Box 4.2 Combinatorial Synthesis

Over the past few decades, the need to discover new drugs has led to development of chemical techniques for producing collections of compounds with various structures, called 'compound libraries', with the expectation that some of these compounds will be identified as drugs with intended pharmacological properties. The techniques of combinatorial chemistry make use of synthetic methods to build large numbers of compounds, basically through a combination of fragments which make up the building blocks. These various fragments are added to the basic building block, or skeleton, which can be based on the core of a molecule with drug-like properties, such as a natural product through the same reaction process, resulting in a group of similar derivatives, say, n_1 in number. Each member of the group is subjected to the next round of reactions with, say, n_2 reactants, resulting in a collection of derivatives. The combination of the two reactions will give rise to a total of $n_1 \times n_2$ products. A library containing a large number of derivatives can be built after only a few rounds of reactions. A simplified scheme of combinatorial synthesis is given in Fig. 4.2.

The generation of a combinatorial library can be made through a series of reactions. In the 'split mix' method, based on solid-phase peptide synthesis, the beads are divided into 20 portions, each undergoing a reaction with a different amino acid. A bead will have a product with a unique structural sequence attached. After release of the mixture from the beads, the product pool can be tested for the activity required, for example, as an enzyme inhibitor or as an

(Continued)

Box 4.2 *(Continued)*

antimicrobial compound. The identity of the active component can be found through a strategy called deconvolution. In the 'parallel synthesis' method, pioneered by Geysen [65], the library members are separated as compounds attached to pins, which undergo parallel reactions in a multiwell plate producing a known unique sequence for each one.

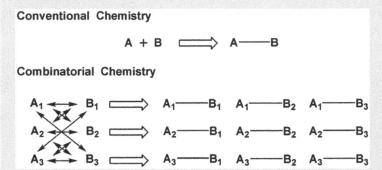

Figure 4.2 A simplified scheme of combinatorial chemistry as compared to conventional chemistry. The starting point can be a template molecule, say, a natural product, which can be made to undergo a reaction with a number of reactants (three in this case). Their further reaction with another three reactants will yield nine products, which can be identified and selected.

Often, however, combinatorial synthesis only generates compounds with similar structures which occupy only a small fraction of the total possible chemical space, and optimal drug candidates may be missed. It has been noted that only one new drug, an antitumour compound sorafenib, has emerged from combinatorial chemistry over the preceding three decades since its development [20]. It has, however, been very useful in optimizing structures of known drugs for better efficacy and safety. Natural products have been useful in providing starting structures for optimization. A technique called dynamic combinatorial chemistry [66] is also potentially useful in making new drug candidates. It makes use of reversible reactions between various structurally similar components which can combine reversibly on a molecular target, serving as a 'mould' to stabilize the compounds with the best binding affinities and hence provide the favoured products.

Other promising approaches are available to enhance the diversity of the synthesized molecules and can often be used together with combinatorial synthesis. One such approach, named diversity-oriented synthesis [67, 68], allows diversity to be incorporated into the library of synthesized compounds. Structural diversity is derived from four components, namely, diversity of the basic skeleton, of the various building blocks attached to the skeleton, of the various functional groups and of the stereochemical (three-dimensional) orientation. As in the case of combinatorial synthesis, natural products offer very good starting points through provision of diverse basic skeletons, from which other components of diversity can be added. Incorporation of these various aspects of diversity in the synthesis schemes give rise to libraries which occupy a much larger chemical space and thus much larger diversity than conventional combinatorial libraries. An example is the development of a new class of antimalarial drugs, the spiroindolones [44, 69]. First, random screening of libraries containing 10,000 synthetic compounds and 2000 natural products gave 275 compounds with good activities. One of these was singled out for further development through diversity-oriented synthesis, giving rise to drugs of the spiroindolone family, which turned out to be a very active class of antimalarial drugs.

The need to make large numbers of compounds in various chemical libraries was met by new developments in organic synthesis, such as 'click chemistry' and metathesis. Some of these developments are briefly explained in Box 4.3.

Box 4.3 Recent Developments in Organic Synthesis to Meet the Need for New Drug Development

Drug development, both from target-based design or from random screening of libraries of compounds, requires the expertise of synthetic organic chemists in producing drug candidates or libraries of potential compounds. The need is met both by conventional synthetic methods and recent ones based on new concepts, often with the help of new catalysts [81].

A concept developed by Nobel laureate Barry Sharpless and associates, in what is commonly called click chemistry, achieves facile synthesis by using reactions which are energetically favoured so that the reaction proceeds easily as if by clicking [82, 83]. The most

(Continued)

Box 4.3 (*Continued*)

important example of such reactions is the copper-catalysed reaction between an azide and an alkyne, both high-energy components, to give 1,2,3-triazole in high yields (see Fig. 4.3a). An advantage of this click reaction is that it can proceed in water so that it can be used to make biologically active compounds as drug candidates. An elegant example of the use of click chemistry in making new drugs is to perform the reaction in the presence of the molecular target, which can act as a template for selecting the components of the best inhibitors which come together and combine in the click reaction at the active site. This was shown for the synthesis of very high-affinity acetylcholinesterase inhibitors from arrays of potential building blocks [84, 85].

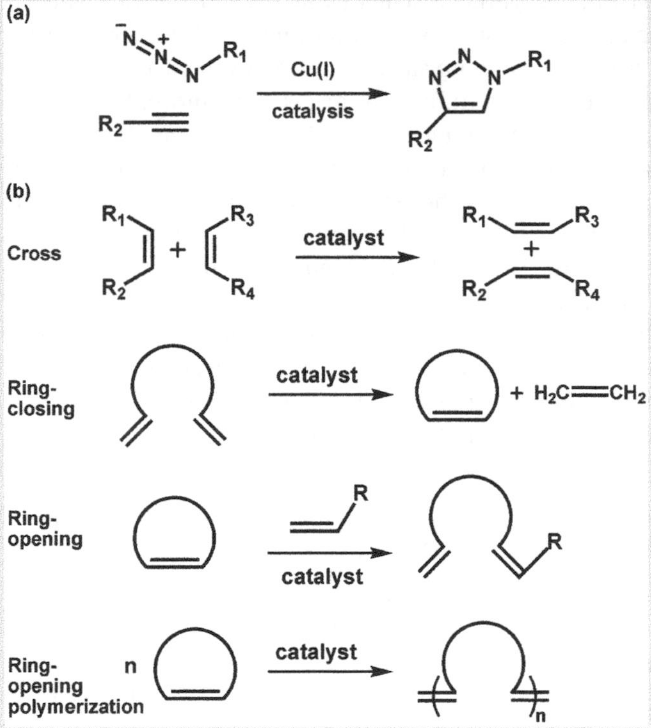

Figure 4.3   (a) Formation of 1,2,3-triazole from a click reaction between an alkyne and an azide. Click reactions can be performed in the active site of the molecular target so as to give rise to high-affinity inhibitors from an array of potential reactants. (b) Some reaction schemes for olefin metathesis, giving rise to exchange between different olefins and new molecular structures.

Box 4.3   (*Continued*)

Another important development, olefin metathesis, is pioneered by Nobel laureates Yves Chauvin, Robert H. Grubbs, and Richard R. Schrock [86, 87]. This is an organic reaction catalysed by metal complexes, which gives rise to new alkenes through redistribution of alkene fragments of the reacting components (see Fig. 4.3b). It is now used quite extensively not only in pharmaceutical chemistry but also in petrochemical industries for the formation of new olefin compounds.

These and other organic synthesis methods are very useful for making new compounds by design or making libraries of compounds. Many of these compounds start from templates of natural products and hence extend the application of their use in the pharmaceutical industry.

4.4   Selection of Drug-Like Molecules: General Molecular Characters for 'Druggability'

Libraries of diverse molecules, both based on natural products and on other chemical structures, form the starting point for drug discovery. These libraries can be screened through high-throughput assays and selected as hits on the basis of their effects on target organisms, tissues or molecular targets, as described in Chapter 3. This is followed by rounds of optimization—modification of the structures to obtain greater efficacy and selectivity—until lead drug candidates are obtained.

Eventually some, albeit very few, drug candidates emerge from such a stringent selection process, at the rate of about 1 in 10,000. The majority of compounds are not druggable, that is, they do not meet the criteria of drug-like molecules in being able to reach the cells and in many cases cross membrane barriers to access the targets. Successful candidates must also remain long enough in the body to be able to exert their action specifically, with minimal toxic or unwanted side effects. To increase the chance of success, some generalizations have been made on characters of drug-like molecules. Those which fail generally have high lipophilicity and low water solubility, that is, they are too oily. From extensive observation to help improve the probability of success of compounds to progress from positive primary activity assays to becoming drug candidates.

Lipinski [70] proposed a 'rule of five' to guide selection of drug-like molecules. This rule states that oral drugs usually have the following properties: a molecular weight of no more than 500, a log *P* (lipophilicity, measured by the ratio of solubility in octanol and water) of less than 5, a maximum of 5 hydrogen bond donors, and a maximum of 10 hydrogen bond acceptors. Another important observation, attributed to Veber [71], is that compounds with good oral bioavailability tend to have fewer than 10 rotatable bonds and a moderate polar surface area, with 12 or fewer hydrogen bond donors and acceptors. These rules, based on observation, are generally useful, although there are exceptions such as compound classes which are substrates for biological transporters as well as many natural products.

4.5 Fragment-Based Drug Discovery

An alternative approach to high-throughput screening for obtaining drug candidates is fragment-based drug discovery.[72, 73] Rather than trying to find whole molecules with good affinity for a specific target, this approach identifies small fragments, that is, small molecules which although binding weakly to the target do so with high efficiency on a per atom basis. Fragment library hits constitute starting points for creating larger-molecule drug candidates. A fragment library has an advantage over full-size compound libraries in that it is much smaller in size but with the potential to be grown into larger collections of compounds with more parts added from design. Design of a fragment library and further elaboration of the fragments make use of structural information of the active site of the target.

Fragment-based drug discovery therefore takes advantage of both library screening and structure-based design. A fragment derived from library screening may initially have only weak binding due to its small size, but the final molecule derived from it and other fragments may have very high binding affinity and potent activity due to the structure-based design. Elaboration of a fragment to form complete molecules may be done by growing the fragment through addition of more parts. Alternatively, it may be done by linking two or more fragments together or merging the fragments through their overlapping parts. Like combinatorial and diversity-oriented libraries, the synthesis of compounds in a fragment-based library is achieved both by conventional and by new methods in organic chemistry. These new methods, some of which are outlined in Box 4.3, are applicable both for making designed drugs as well as for making diverse libraries of compounds for screening purposes.

Natural products are rich in fragments which have structural diversity, and therefore offer very good scaffolds for fragments in the making of diversified libraries [74], which can be screened for potential drug molecules.

4.6   Expanding Drug Diversity through Biology

Antibiotics and other drugs from nature are made from series of reactions, many of which are those in what is called secondary metabolism, namely, metabolism which is not the central part of an organism's life processes. Secondary metabolite compounds provide a survival or growth advantage, such as attack or defence against other organisms. By manipulation of the genes which are involved in synthesis of these metabolites, new compounds may be formed, which might be called 'unnatural natural products', some of which can become new drugs [75, 76].

As examples of this approach to producing new drugs by biosynthesis, we might consider the polyketides, which occur naturally in a variety of organisms, including bacteria, fungi, plants and animals. Some polyketides, including erythromycin, tetracycline and rifamycin, display antibiotic activities, while others such as amphotericin have antifungal properties, and yet others such as doxorubicin and geldanamycin have antitumour activities. Polyketides are synthesized by a group of enzymes known as polyketide synthases, which have different specificities for various precursors. By manipulating the synthesis of polyketides through introduction of polyketide synthase genes into microorganisms, they can be programmed to produce new polyketides which are not found in nature [75]. Indeed, by inserting a variety of polyketide synthases in a microorganism, a variety of polyketides can be synthesized through different combinations of a limited number of reactions, a process called combinatorial biosynthesis. This is a promising way to produce diverse natural compounds which are not found in nature.

The conventional way of discovering new drugs from nature through cultivating microorganisms and harvesting their products can be supplemented by the techniques of metagenomics [30]. The DNA of many microorganisms, from the soil, the sea or other sources, which cannot be grown artificially, can be extracted from nature and put into microorganisms. These can then be grown and manipulated to yield the products, which can be tested for therapeutic and other properties [31, 77]. This is a promising method for obtaining new drugs, such as antibiotics effective against microbes which have become resistant to conventional antibiotics. It is also a good way to search for genes responsible for antibiotic resistance [78, 79].

Another way to generate diversity through biology is to identify metabolites of compounds which may be transformed into active drugs in the process of its metabolism. Some examples are known from the field of antimalarial chemotherapy, where compounds such as desethylchloroquine, a metabolite of chloroquine, and Feb-A and Isofeb-A, metabolites of febrifugine, have been shown to have potent antimalarial activities [80].

4.7   Need for New Drugs in the Pipeline

The process of generating chemical diversity and finding new drugs against infectious and other diseases has to be an ongoing process. Recurring problems of drug resistance, or loss of drug effectiveness due to various reasons, necessitate continuous search for new effective drugs. The pipeline of new drugs has to be filled so that those which have lost effectiveness can be replaced. The problems can be viewed as ongoing attempts to deal with molecular wilderness with its ability to adjust and restore the balance. We can tame the wilderness manifested in diseases but only temporarily in most cases. Nature finds ways to fight back and gain lost ground, necessitating our search for new remedies, as will be discussed next.

Chapter 5

The Wilderness Fights Back

The truth, seldom mentioned but there for anyone to see, is that nature is not so easily molded.

—Rachel Carson, *Silent Spring*, 1962

Summary

Microorganisms and diseased cells can develop resistance to drugs through various mechanisms. Drug resistance can be viewed as the natural tendency of the wilderness to fight back against human intervention, just as life evolves from the struggle for existence in the natural world. Poor human behavior and public health practices contribute to the emergence of drug resistance. We need to understand the mechanisms of drug resistance and find rational approaches to overcome the resistance, either by modifying old drugs or by finding new drugs, including drug combinations. Good examples of natural combinations can be found from the strategies which microorganisms use to prey on others or defend themselves in the ecosystem. In addition to the emergence of drug resistance, new, emerging diseases resulting from global climate and social changes also add to the threats of the molecular wilderness, requiring vigilance and quick responses, coordinated on a global scale.

Tapping Molecular Wilderness: Drugs from Chemistry–Biology–Biodiversity Interface
Yongyuth Yuthavong
Copyright © 2016 Pan Stanford Publishing Pte. Ltd.
ISBN 978-981-4613-59-0 (Hardcover), 978-981-4316-60-6 (eBook)
www.panstanford.com

5.1 'Life Finds a Way'

The novel and film *Jurassic Park* tells the story of bringing dinosaurs back to life from their DNA preserved in amber. Even though a technology was designed to prevent them from breeding, nature helped them to breed and spread. 'Life found a way', the scientists concluded.

Although the story is a fiction, parallels of life finding ways to break free and spread are found everywhere. Indeed, the theory of evolution by Charles Darwin and Alfred Russell Wallace puts the struggle for existence and survival of the fittest as the cornerstone for the evolution of life forms. Evolution occurs in nature from interactions between life forms and between life forms and their environment. Human beings contribute new factors influencing the course of evolution through interferences which are intended to promote their own species, with unintended consequences for evolution.

We use antibiotics to fight infectious diseases by killing the germs which cause them. We use insecticides and pesticides to kill insects and other pests which harm us or which make our lives inconvenient. For a while these tools work in our favour, but then, sooner or later, these threatened life forms will evolve ways to fight back against us. Nature has many ways of creating diversity of organisms against which we aim our weapons. A few members of the species under our attack can resist them, and these will proliferate and go on to become even more resistant through various mechanisms. In our battle against disease pathogens with antibiotics, we soon encountered resistance and had to modify our strategies, for example, by using different drugs or by using drug combinations. This would work for some time but not permanently. A new antibiotic or its combination with other drugs would work for a period of say 5 to 10 years, after which resistance would grow and new agents would be needed. The situation is akin to a war, with an ongoing arms race, in which one side increases its weapons capability, only to be retaliated by the other side, which can also increase its capability to fight back. This ongoing war means that we need to understand the mechanisms of drug resistance and find new ways to deal with them or, better still, to prevent them from occurring altogether.

5.2 Drug Resistance: A Problem of Increasing Urgency

Drug resistance is an increase in the capability of pathogens or diseased cells, such as cancer cells, to withstand the effects of the drugs which are intended to kill them or inhibit their vital functions. Resistance can also occur with drugs used to control our physiology, such as insulin used to treat diabetes by lowering the blood glucose level. As a result of resistance, the dose of the drugs must be increased in order to obtain the same effect or new drugs have to be deployed. A related term, 'drug tolerance', is used to indicate diminished response of the host, such as the patient, to a drug, which may occur through physiological changes in the host, such as increased drug metabolism.

Drug resistance, especially resistance to antimicrobial drugs, is a problem of increasing urgency [88]. Some examples illustrate the scale of this problem:

- Multidrug-resistant tuberculosis (MDR-TB), with resistance to rifampicin, isoniazid and other drugs, is a widespread global problem making therapy much more difficult, lengthy and costly. The World Health Organization [89] has estimated that there are about 630,000 MDR-TB cases in the world.
- Malaria resistant to chloroquine and pyrimethamine-sulfadoxine arose and spread from Southeast Asia to the rest of the tropical world. Resistance to the current artemisinin family of drugs is also now established.
- Methicillin-resistant *Staphylococcus aureus* (MRSA) is often encountered in hospital-acquired infection and infection of wounds and the bloodstream from other sources. Other serious resistance problems are found for *Escherichia coli* and *Klebsiella pneumoniae* causing urinary tract, bloodstream and other infections.
- HIV/AIDS resistance to such drugs as reverse transcriptase inhibitors and protease inhibitors is common.

Antimicrobial resistance occurs relatively easily due to many mechanisms, as will be described next. The fast propagation of microorganisms allows these mechanisms to spread resistance effectively. This, together with our lack of good practice in using drugs, leads to the emergence and spread of antimicrobial resistance.

5.3 Mechanisms of Drug Resistance

There are many mechanisms for drug resistance. As explained earlier, for example, in Chapter 3, a drug can exert its effect by acting on its molecular target. It binds specifically to the target, resulting in inhibition or modification of the function of the target. A common mechanism for drug resistance is alteration of the affinity of the target molecule for the drug, resulting in decreased binding and consequently decreased inhibition or alteration of the function of the target. This alteration can occur through mutations, which can introduce factors which interfere with normal binding of the drug to the target. One common factor is introduction of a mutation which modifies the target so as to reduce the binding affinity of the drug through changes in space availability or in chemical characters such as polarity or charge distribution in the active site. For example, the resistance mutation may replace the original side chain of an amino acid in the active site with a more bulky one, causing steric (spatial) conflict with the drug and hence reducing its binding affinity. Box 5.1 gives an example of this mechanism of resistance found in the malarial parasite *Plasmodium falciparum* against the antimalarial drug pyrimethamine, which normally kills the parasite by inhibiting its dihydrofolate reductase (DHFR) enzyme, a key enzyme in DNA synthesis and other metabolic pathways. In the resistant parasites, mutations reduce the binding of pyrimethamine to this enzyme and hence the ability of the drug to kill the parasites.

Another mechanism of drug resistance is increase in the quantity of the target molecules so as to overwhelm the effect of the drug, which can bind only to a limited number of target molecules. An example of this mechanism is resistance to the cancer drug methotrexate, which acts by inhibiting human DHFR, stopping DNA synthesis and other metabolic changes linked to malignancy. Methotrexate-resistant cancer cells can overwhelm the effect of the drug by increasing the enzyme production.

Other mechanisms of drug resistance involve destruction or inactivation of the drug. For example, antibiotics of the penicillin family, which rely on the integrity of the beta-lactam function, can be inactivated with the destruction of this function. Many pathogenic bacteria develop resistance to these drugs by acquisition of beta-lactamase enzyme. Others change their target protein for penicillin and penicillin-binding protein (PBP) so that the drug can no longer bind and exert its effect.

Reduced entry of the drug is another mechanism, found in bacteria which are resistant to aminoglycoside antibiotics, among others. Many drug-resistant pathogens can cause increased efflux of the drug through operation of efflux pumps. An important protein, P-glycoprotein, is widely distributed in various cells and can pump out many drugs and toxic molecules through an ATP-fueled mechanism. This physiological mechanism has been adopted by many pathogens and diseased cells to provide resistance to various drugs, called multidrug resistance (MDR). Figure 5.1 gives a simplified diagram of various mechanisms of drug resistance.

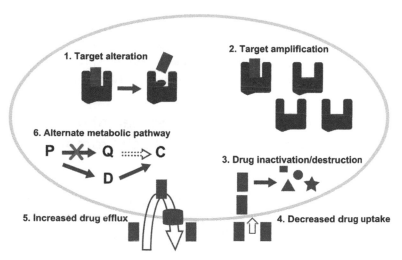

Figure 5.1   Mechanisms for drug resistance. Some mechanisms are based on changes in the molecular target, such as (1) reduced drug binding or (2) increase in the number of target molecules. Some are based on (3) drug inactivation or destruction, (4) reduced drug uptake or (5) increased drug efflux from the target cell (e.g., pathogen, cancer cell), while others are based on (6) an alternate metabolic pathway which the cell can use to bypass the blockage of the original metabolic route by the drug.

How do the pathogens or diseased cells acquire the capacity to become resistant?

In some cases, this occurs from natural variations in the genetic makeup of the cells, manifested as mutations directly in the target genes, or elsewhere, leading to variation in the control of gene expression. These variant cell types, or mutants, are more capable of withstanding drug pressure, and they are selected for when the drugs are deployed against them; the less hardy variants, including the majority 'wild type' in the population, die out. In some cases, the genes of the drug targets are amplified, enabling higher expression, while in other cases the expression of the enzymes for alternate pathways is enhanced so as to enable the cells to avoid the effects of the drugs.

In many cases of drug resistance in microorganisms, the resistance genes are acquired from outside sources through plasmids, small circular DNAs which encode the genes for drug inactivation, for pumping out the drugs or for some other mechanisms which enable resistance. The plasmids and other genetic elements which enable drug resistance act like currencies which can be transferred from one microorganism to another, and help to explain the facile emergence of drug resistance in a variety of microorganisms.

5.4 Ease of Occurrence and Spread of Drug Resistance and Factors Which Promote Them

Occurrence of resistance of pathogens to drugs such as antibiotics is an example of natural selection, which can occur quickly because of relatively fast reproduction of the pathogens and their large numbers in infection. Not only do natural variations in drug sensitivity due to variations in the genes of drug targets already exist in nature, but new mutations also arise continuously with replication of DNA. Take the case of resistance due to mutation in a target as an example. Although random mutation occurs relatively rarely, say one in 10^7 bases per generation, such fast reproduction and large numbers of the pathogens ensure that there will always be significant numbers of new mutants, from which the selection for drug resistance will be achieved by the pathogens. Acquisition of factors for drug resistance, such as resistance plasmids, is also a relatively common event, since bacteria and other microorganisms can be transformed, that is, take up external DNA relatively easily, and they possess a rudimentary sexual process called conjugation, which drives the exchange of genetic materials from one microbial cell to another.

A common approach to combatting drug resistance is to increase the dosage of the drug. However, this is in most cases only a short-term solution and leads to even higher resistance due to accumulation of mutations, gene amplification or other mechanisms, which together increase the capability of the pathogen to withstand the effect of the increased dosage.

Given the ease at which antimicrobial drug resistance can evolve, the way we have been using these drugs has inadvertently helped in the emergence and spread of resistance. Some of the factors which lead to drug resistance [89] include:

- Lack of a comprehensive and coordinated response of the public health system to the emergence of drug resistance. Levels of resistance can increase and resistant pathogens can spread to wider locations in the absence of such a response.
- Weak or absent antimicrobial resistance surveillance and monitoring systems. Without appropriate surveillance, emergence of resistance cannot be checked and stemmed.
- Inadequate systems to ensure quality and uninterrupted supply of medicines. Low-quality and insufficient dosages allow resistant pathogens to survive and the levels of resistance to rise.
- Insufficient diagnostic, prevention and therapeutic tools. Lack of adequate tools and their use allow resistant pathogens to become established and spread.
- Inappropriate use of antimicrobial medicines, including in animal husbandry where antibiotics are used to enhance animal growth, thereby leaving residual amounts in the animals and releasing excretions to the environment. Such residual amounts, both in the animals and in the environment, encourage development of resistance of the pathogens.

Antimicrobial drug resistance has now become a problem of such magnitude that it has been predicted that 'the golden age of antibiotics' could come to an end unless serious action is taken [90].

The problems require not only individual efforts but also good public health policy to ensure optimal drug use and to minimize the risk of emergence of drug resistance. At the same time, we need to strengthen the existing tools and find new scientific solutions to tackle the problems of drug resistance.

5.5   Tools to Fight Drug Resistance: New Drugs

While a number of factors contributing to drug resistance emergence and spread can be controlled by better public health policy at local, national and international levels, we should seek essential tools from science in order to fight this problem. Using drugs at higher doses against which resistance has already taken hold would only provide a temporary solution to the problem until resistance inevitably rises to higher levels. We need therefore to find more effective solutions.

In this effort, it helps to recall the rules of engagement of molecular wilderness. In the complex world of interacting organisms, each individual species has its own molecular repertoire to help in its survival and to defeat its enemies. Antibiotics, for example, are molecules produced by microorganisms and other living organisms to kill their enemies or prey. In return, the enemies or prey, which are often other microorganisms, can find ways to avert or neutralize the deadly effects of the molecular weapons aimed at them. The strategies to survive in the molecular wilderness of nature have been adopted by drug-resistant pathogens, since the mechanisms used by them to survive in nature can also be used to fight man-made arsenals in the form of drugs. To deal with drug resistance, we should have multiple strategies, with the anticipation that the pathogens or diseased cells may sooner or later find ways to get around one or more of them.

An obvious strategy is to find new effective molecules. There are two main ways of doing this. One, a conservative way, is to find new derivatives of the old drug which are effective, although there is resistance to the old drug. The new derivatives still retain affinity to the drug target owing to differences in structures, such that new and old drugs bind in subtly different ways. We have seen this strategy at work in Chapter 4, Box 4.1, where many antibiotics are derivatives of penicillin or have structures containing beta-lactam functions. Some semisynthetic penicillins, such as cloxacillin, are effective because their lactam functions are resistant to the lactamase action of the penicillin-resistant microorganisms.

An example of the development of new effective molecules which could neutralize the effect of resistance mutations is shown in Box 5.1. In this case, the molecular structure of the mutated target, the malarial DHFR, was determined to reveal the main cause of resistance, which is the inability of the old drugs to bind the target due to obstruction created by the mutations. New molecules were designed to avoid the obstructions and allow binding to the target, resulting in inhibition of the enzyme and restoration of drug efficacy. This is an example of a successful approach to deal with resistance.

Another way of finding new effective molecules is to discover new molecular entities which work in ways different from the earlier ones against which resistance has arisen. This can be done by two main approaches. One approach is rational design of compounds against different targets from those which are known already to encounter resistance. An example of success by this approach is the discovery of inhibitors of human immunodeficiency virus (HIV) integrase, an enzyme which integrates HIV genetic material into host cell DNA, which forms a new family of HIV drugs distinct from the reverse transcriptase inhibitors and protease inhibitors [91, 92]. Another approach is to screen compounds of various structures, either from chemical libraries obtained in various ways outlined in Chapter 4, for example, or obtained from natural extracts of plants or microorganisms, against drug-resistant microorganisms or diseased cells. Active compounds can be studied further in order to gain more information on their targets and mechanisms of action. An example of this approach is the discovery of spiroindolones as new antimalarials, which act through inhibition of the malaria parasite's sodium transport [44].

5.6 Tools to Fight Drug Resistance and Improve Efficacy of Existing Drugs: Drug Combinations

A proven way to counter drug-resistant pathogens or diseased cells is by combining different compounds together in a treatment [93,

Box 5.1 Mechanism of Resistance of the Malaria Parasites against Pyrimethamine and Mechanism-Based Development of New Effective Drugs

Pyrimethamine and its combination with other drugs have long been used to treat falciparum malaria, the severe form of malaria caused by the malaria parasite *P. falciparum*. Its molecular target is the enzyme DHFR, used by the parasite in a crucial step for its synthesis of DNA and other essential metabolites (for background information, see Box 3.1, Chapter 3). However, resistance arose against pyrimethamine from mutation at residue 108 of DHFR, changing from serine to asparagine, and subsequently at other residues, with associated loss of binding affinities [97, 98]. The three-dimensional structure of DHFR from resistant parasites shows that the mutation at residue 108 introduced a bulky side chain, which interferes sterically with the chlorine atom of pyrimethamine (see Fig. 5.2a) [49]. Consequently, the efficacy of the drug is decreased. Additional mutations lowered the affinity of the drug for DHFR further and increased the resistance of the parasite. A structurally similar drug, cycloguanil, also encountered a similar mechanism of resistance, compromising its usefulness.

Knowledge of the mechanism of resistance helped in the designing and making of new effective drugs. It was earlier shown that the side chain of cycloguanil, with the sterically hindering chlorophenyl group, could be replaced by a flexible group which enabled high-affinity binding with retention of drug efficacy [99, 100]. The side chain of pyrimethamine was likewise replaced with similar beneficial effects and the additional benefit of high oral bioavailability (i.e., the drugs can be taken orally). These compounds were shown to bind with the active site of mutant DHFR with high affinity (Fig. 5.2b). Addition of fragments which could bind with other parts of the active site strengthened the binding affinity to parasite DHFR relative to human DHFR, hence enhancing the selectivity of the drug. The mode of binding of P218, one of the best compounds of the series [101], is shown in Fig. 5.2c. The side chain could avert the steric conflict with asparagine (residue 108) and also reach out to bind with arginine (residue 122) and other parts of the active site, in contrast with the active site of human DHFR, where P218 cannot make such extensive interactions. The drug is selective and highly efficacious against both wild-type and pyrimethamine-resistant parasites.

(Continued)

Box 5.1   (*Continued*)

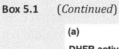 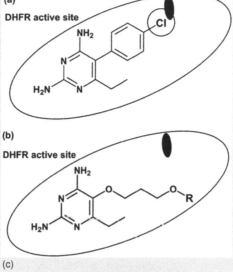

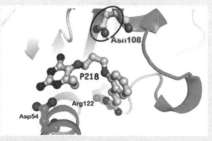

Figure 5.2   (a) Resistance of malarial parasites (*P. falciparum*) to the antimalarial pyrimethamine occurred through mutations of the drug target, the enzyme dihydrofolate redustase (DHFR). A mutation from serine to asparagine in the active site introduced extra atoms (represented by solid oval), which make a steric conflict with the chlorine atom (circle) of pyrimethamine, reducing its affinity for the enzyme, which is subsequently reduced even further by additional mutations. (b) New effective inhibitors against the parasites can avoid the steric conflict and retain binding affinity with the mutant enzymes. (c) P218, one such new inhibitor, in the mutant DHFR active site, with the mutated site causes steric hindrance to pyrimethamine but is averted by the new inhibitor in the circle. Picture courtesy of Dr Penchit Chitnumsub, BIOTEC Thailand.

94]. The combination can include different drugs or drugs paired with compounds with the capacity to fight resistance but not necessarily drugs by themselves. The drugs can be chosen on the basis of actions on different targets or on different mechanisms of action. For example, if spontaneous drug resistance of a target occurs with a frequency of say 10^{-7} in using a combination of two drugs acting on different targets, the probability of resistance occurring from both targets is $10^{-7} \times 10^{-7}$, that is, 10^{-14}. Similar considerations can be made even for drugs which inhibit the same target but at different sites or with a different mechanism. Drug combinations can also be synergistic, giving a total therapeutic effect greater than the sum of the effect given by each separately.

Drug combinations can be roughly divided into three categories, namely, those which inhibit targets in different metabolic pathways, those which inhibit targets in the same pathway and those which inhibit the same target in different ways, such as binding at different sites (Fig. 5.3). Combinations of drugs which inhibit targets in different pathways have been used successfully to treat infectious diseases, cancer and other diseases in which drug resistance has arisen. For example, resistance to a tuberculosis drug given on its own arises so quickly as to render it incapable of curing the patient. Therefore, effective treatment now requires drug combinations. The regimen called directly observed treatment, short course (DOTS) employs a combination of isoniazid (a fatty acid synthase inhibitor), rifampicin (an RNA polymerase inhibitor), ethambutal (a cell wall synthesis inhibitor) and pyrazinamide (a drug with a mechanism of action not yet well understood). HIV infections used to be untreatable resulting in high mortalities as the virus could quickly develop resistance to single antiviral drugs. Nowadays, HIV can be contained as a chronic infection with treatment based on combinations of a mixture of two nucleoside reverse transcriptase inhibitors (emtricitabine and tenofovir) with raltegravir (an integrase inhibitor), efavirenz (nonnucleoside reverse trancriptase inhibitor) or a mixture of ritonavir and darunavir (both protease inhibitors). Combination therapies are now also used in treatment of malaria, which developed resistance against single drugs such as chloroquine and pyrimethamine. In these combinations, a derivative of artemisinin is given together with other drugs such as mefloquine, amodiaquine or lumifantrine. These drugs likely act through different mechanisms, although their targets of action have not been clearly identified.

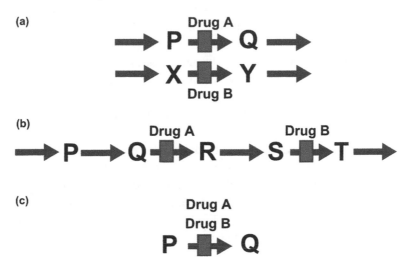

Figure 5.3   Three main categories of drug combinations, exemplified by a combination of two drugs A and B, which (a) inhibit targets in different metabolic pathways, (b) inhibit targets in the same pathway and (c) inhibit the same target in different ways.

In some cases, a partner in the combination may not have a killing effect but augment the action of a drug such as a lactam antibiotic which may otherwise be compromised by the resistance mechanism, such as the action of beta-lactamase. For example, clavulanic acid is a beta-lactamase inhibitor which has been combined with amoxicillin, an antibiotic in the penicillin family, in the oral formulation named Augmentin used against many otherwise resistant bacterial infections. Another example is employment of agents which can reverse the action of drug efflux transporters responsible for MDR in cancer and other infectious diseases.

Successful combinations of drugs against targets in the same pathway include those which inhibit enzymes in folate cofactor biosynthesis pathways in bacterial and protozoal diseases such as malaria. Drug combinations against folate pathway targets are potentially useful as they have little side effects; this is because humans do not synthesize the folate cofactors but rely instead on salvage from food sources. Moreover, these drug combinations were found to be synergistic, giving a total effect greater than the sum of effect of each partner. Clinically useful combinations are those of inhibitors of dihydropteroate synthase and DHFR (see Box 5.1), such as sulfamethoxazole and trimethoprim (antibacterial combination, marketed as, for example, Bactrim) and sulfadoxine and pyrimethamine (antimalarial combination, marketed as, for example, Fansidar), respectively. Inhibitors of early stages of cell wall synthesis, such as tunicamycin, are also synergistic with lactam antibiotics, which inhibit the late stage of synthesis.

Effective drug combinations can be directed against a single target but at different sites or through different mechanisms. The bacterial ribosome, the machinery for protein synthesis, provides an example of such a target. A combination consisting of a nonribosomal peptide and a polyketide–nonribosomal peptide hybrid, which bind in adjacent sites of the 50S subunit of the ribosome, provide an effective regimen, which is much more potent than either component alone.

In spite of the fact that drug combinations can help to fight the problem of drug resistance, they need to be deployed carefully before resistance occurs to any of the partners in the combinations which could compromise its efficacy. Another important point is that the pharmacokinetics of the partners need to be matched so that during the course of treatment the pathogens or diseased cells are mostly exposed to the combination of all partners.

5.7 Natural Combinations

Nature provides examples of combinations of chemical weapons such as antibiotics, which microorganisms use to prey on others or to defend themselves. We can learn from these natural combinations, which must have evolved in order to solve the problems of resistance to single-chemical compounds. Indeed, successful use of natural product concoctions for remedies in traditional medicine may be akin to the successful natural combinations used by many microorganisms. Resistance to antibiotics has evolved in nature due to need for survival of target microorganisms and a state of balance is struck between them and the antibiotic producers [95, 96]. To have effective chemical weapons, antibiotic-producing microorganisms often produce a large number of antibiotics, combinations of which are effective against their enemies which may be resistant to single antibiotics in the mixture [93]. For example, a bacterial strain can have multiple gene clusters producing a large number of natural products which together form effective combinations against its enemies. A strain of *Streptomyces* (*S. clavuligerus*) produces the beta-lactam antibiotic cephamycin together with the beta-lactamase inhibitor clavulanate – a strategy later used in clinical treatment of resistant bacterial infection. Study of these natural combinations should lead to other effective clinical drug combinations in the future.

5.8 Emerging and Re-Emerging Diseases

The molecular wilderness is not static but evolving continuously in response to the environment, as we see from examples of drug resistance arising from our use of drugs. In a broader context, the nature and variety of diseases affecting us are continuously changing due to a number of factors, ranging from changes in the environment, climate, living conditions and other social and economic factors. Typically, an emerging disease is caused by a pathogen such as a virus originally found in animals, that is, a zoonotic disease. AIDS is caused by HIV, believed to have originated from nonhuman primates living in sub-Saharan Africa. Some emerging diseases arise originally from poultry, such as bird influenza, aided by poor sanitary conditions, ease of travel of humans and natural migration of birds. Ebola viruses which caused outbreaks in West and Central Africa are probably first transmitted from wild animals and are spread by human-to-human transmission. The outbreak of severe acute respiratory syndrome (SARS) in 2003 was caused by a coronavirus which probably originated in bats and in other animals, including civets, and more recently another coronavirus likely from camels was responsible for the Middle East respiratory syndrome (MERS) in 2013. Emerging diseases are often caused by viruses, which can potentially jump from one host species to another and can develop resistance to drug treatment due to their fast genetic changes.

Some diseases which appeared to have been suppressed can re-emerge to become threats to various communities. In some cases, this is due to drug resistance problems, which have become widespread, such as MRSA infection or MDR-TB. In other cases, old diseases such as plague or cholera re-emerge in communities with poor living conditions or under pressure of war or starvation. Global environmental changes, such as climate change, can also significantly influence the incidence of diseases. For example, mosquito populations can explode with changes in temperature or rainfall, resulting in an increase of mosquito-borne diseases.

These emerging and re-emerging diseases are clear reminders of the threats originating from nature, and they require continuous vigilance and response. New drugs and drug combinations are important components of our response, but they form only a part of the big picture on how we can live with the molecular wilderness. We need continuous surveillance of disease occurrence, which is aided by simple and accurate diagnosis. The environment and living conditions of people also need to be continuously monitored with regard to their influence on the emergence or re-emergence of diseases. Preventive measures include not only vaccines, available for only some diseases and often at a cost too high to be deployed to the majority of the population in many developing countries, but also various actions at the individual and public health levels in line with good practices for maintaining health. These practices concern not only the appropriate use of drugs, vaccines and diagnostics but also broader aspects of maintaining or improving our relation with the environment and achieving sustainability of the global community, not only us, but also the whole wilderness of nature.

Chapter 6

Living with Molecular Wilderness

Wilderness is not a luxury, but a necessity of the human spirit, . . .
—Edward Abbey, *Desert Solitaire: A Season in the Wilderness*, 1968

Summary

Learning from past lessons, we should come to realize the power of molecular wilderness, both to yield benefits for the human species and to strike back when we oversimplify its exploitation and disregard the delicate balance of nature. Coexistence and conservation should be preferred over exploitation and subjugation of the molecular wilderness. Sustainable tapping of the molecular wilderness requires not only science and technology but also a balanced approach, taking into account the social, economic and environmental factors affecting the health of people all over the world. Furthermore, since biodiversity is mostly endowed in tropical countries, where the standards of health care are still poor, sustainable tapping of the molecular wilderness should also be done with the objective of improving these standards so as to achieve a healthy world for all.

Tapping Molecular Wilderness: Drugs from Chemistry–Biology–Biodiversity Interface
Yongyuth Yuthavong
Copyright © 2016 Pan Stanford Publishing Pte. Ltd.
ISBN 978-981-4613-59-0 (Hardcover), 978-981-4316-60-6 (eBook)
www.panstanford.com

6.1   Lessons from Molecular Wilderness

Just as a person learns from his or her successes and mistakes, so should we as the human society learn from the past in our dealing with the wilderness so as to build from successes and correct mistakes. In the past, the molecular wilderness has given us innumerable raw materials, to which science and technology have added value and have used its products as drugs to fight diseases and other ailments, not to mention numerous nonmedical uses. As we have seen, however, these diseases and ailments are also themselves the products of the molecular wilderness, and they can find ways to fight back. The past approach has been temporarily successful, but there are now serious doubts over its sustainability.

Take the case of antibiotics. Over the past few decades, we obtained many antibiotics from nature, which we have used to treat our illnesses, at first with miraculous success. We have modified the structures of the molecules from nature, figured out industrial methods to make them in large quantities and used them as therapies for infectious and other diseases. Advances in medical science on other fronts also reinforced our confidence. Some even dreamt that we were on our way to conquer infectious diseases. For example, the surgeon general of the United States William Stewart said in 1967, 'The time has come to close the book on infectious diseases. We have basically wiped out infection in the United States.'

After a few decades of the success of the antibiotics, however, we are now faced with resistance of pathogens which we thought we had conquered and are also threatened by new, emerging infectious diseases. We have seen examples of drug-resistant pathogens and their origins in Chapter 5. The new, emerging diseases include HIV-AIDS, severe acute respiratory syndrome (SARS), the Middle East respiratory syndrome (MERS), avian influenza, and haemorrhagic fever caused by the Ebola virus. The National Institute of Allergy and Infectious Diseases of the United States has listed 13 newly recognized infectious pathogens, five re-emerging infectious pathogens and many more infectious pathogens with bioterrorism potential [102]. The latter are pathogens which may not be common, or may have been tamed in the past, but can be adopted for terrorism purposes, including pathogens causing anthrax, plague, tularemia and smallpox.

We have to come to the conclusion that our tapping of the molecular wilderness, although yielding wonderful results on the discovery and use of molecules from nature to cure our illnesses, may not yield sustainable solutions, since the wilderness can fight back due to plasticity of life forms. Antibiotics, for example, are weapons made by microorganisms to prey on others or to defend themselves. In nature, they are not used as single agents but as parts of strategies for their interaction with their enemies or their prey. The outcome is evolution of all interacting life forms, each struggling for its own survival. In contrast, our tapping of molecules from nature has been done mostly in a simplistic manner, obtaining single drugs such as broad-spectrum antibiotics, which were then used intensively in ways which do not take sufficiently into account the plasticity of infectious organisms and their ability to fight back. Experience over the past few years has now taught us to control the use of drugs in an attempt to achieve sustainability. Even then, we now realize that each drug against infectious pathogens has a limited life time of usefulness before it is compromised by pathogen resistance.

The realization that the 'magic bullet' does not exist should not deter us from finding the best solutions in the therapy of infectious and other diseases. The problems are complex, and solutions based on science and technology alone, powerful though they are, will not be sufficient. We need input from other areas of human knowledge. At this juncture, it may be a good idea to examine other complex problems confronting us and how we are trying to solve them so that we may learn from these efforts.

6.2 Lessons from Climate Change

One of the great global crises in our time is climate change resulting from human activities, mainly use of energy and our industrial, agricultural and household practices [103]. We have used fossil fuels as main sources of energy for an extended period, resulting in emission of carbon dioxide and other greenhouse gases, with global warming as a consequence. Our industrial, agricultural and household practices contribute to global warming, not only from the use of fossil fuels, but also in the production of other greenhouse gases, such as chlorofluorocarbons used as refrigerants and for other purposes and methane from waste. Deforestation has also contributed to global warming since it reduces the absorption of carbon dioxide by plants. Although climate change is a natural process, over the past few decades, it has been increasingly realized that human activities have significantly contributed to it in the form of global warming. Threats from climate change include droughts, floods, sea-level rise and various other hazards. For example, climate change may increase the incidence and severity of infectious diseases due to effects on abundance and adaptive changes in vectors and pathogens. The risks from climate change therefore have to be mitigated, and we must adapt before these problems reach critical levels in the next few decades.

Climate change is an issue which can offer valuable lessons in our future dealing with the molecular wilderness. For example, overreliance on fossil fuels as sources of energy is akin to overreliance on antibiotics as therapies for various ailments. Consequences such as floods, droughts and sea-level rise from climate change were largely unforeseen, although they could have been anticipated had we been more environmentally aware. Similarly, the rise of resistance of pathogens due to our use of drugs could have been anticipated from our knowledge of mechanisms of evolution. Evolution allows adaptation of life forms on being challenged by various stresses: those which cannot overcome the stress will perish, while those which can deal with the stress, for example, through genetic changes leading to drug resistance, will survive and flourish.

In dealing with problems of climate change, the risks of present and future harms must be mitigated. This was done, for example, in the case of phasing out the use of chlorofluorocarbons, which are a cause of the greenhouse effect and contribute to global warming. Reliance on fossil fuels is increasingly reduced through use of solar and wind energies. Increased effort is made to reforest various parts of the world, together with attempts to reduce deforestation. Furthermore, adaptation must be made to the changes brought about by global warming, for example, by management of anticipated floods, droughts and sea-level rise. Effects on disease occurrence and severity, food production and distribution and other effects must also be made through adaptation of our present way of life and livelihood.

Similarly, in dealing with the problems of drug resistance, we need to have both mitigation and adaptation strategies. The risks of resistance can be mitigated by informed policy for drug use and public education. Increased emphasis should be made on improvement of public health and prevention efforts, including sustainable reduction of vectors, increased surveillance of diseases and status of resistance and mass vaccine administration. One way of preventing resistance from occurring is the use of drug combinations, where members of the combination have different targets or modes of action so that the chance of occurrence of resistance is diminished since it needs changes in all the targets (see Chapter 5). However, even after all the mitigation efforts, problems of drug resistance will likely eventually occur, albeit after longer time periods. We need to adapt to the resistant pathogens once they occur, for example, by finding new effective methods to detect and characterize the resistant pathogens, by changing treatment regimes to ensure efficacy, by introducing new drugs with different mechanisms of action or finding new ways to treat disease. We need to change our behaviour and lifestyle in order to avoid exposure to vectors or pathogens. To avoid exposure to mosquitoes, insecticide-treated bednets have been introduced, with very impressive results on the reduction of malaria in Africa. Chicken farming practice had to be changed in Asia and elsewhere to avoid too close contact between people and birds which posed a risk of bird influenza. The practice of eating 'bush' meat, that is, meat from wild animals from the forest, needs to be discouraged, not only because it encroaches on wildlife, but also because it raises the risk of zoonotic diseases.

These mitigation and adaptation efforts require the tools of science and technology to come up with new drugs, diagnostics, vaccines and other means of fighting diseases. However, similar to our efforts in dealing with climate change, they also require a broader change—a transformation—in our society in order to achieve sustainable solutions. This transformation cannot be achieved only by science and technology, although they are important enabling tools. It requires change among individuals and the society as a whole regarding our outlook and values on our relationship with nature. Furthermore, we should not think of winning the battle with infectious diseases only as a medical or a public health problem separated from other problems concerning the livelihood of people, poverty, human behaviour, religion and culture. In short, we need to think of a broad, ecological approach to the problems concerning health and diseases.

6.3 Ecological Approaches to Treatment and Management of Infectious Diseases

Ecology, the study of interactions among organisms and their environment, offers a broad and promising guide to success in our treatment and management of infectious diseases [104–106]. This comes from the realization that health is a complex issue, and it is achieved only by provision of medical services to individuals allied with the presence of a good public health system. Major factors influencing community health include the environment, the economy, how people earn their living, their education levels, what they eat and how they interact with one another. For example, people living in forest areas earn their livelihood by clearing the forest and planting corn. They are thereby exposed to mosquitoes and are at risk of malaria. Disadvantaged by poverty and a lack of education, they seek relief by medicines bought from local drugstores, which have poor quality and contribute to the development of drug resistance. In another example, people are fond of eating raw freshwater fish, which harbor flukes, which become lodged in the bile duct, causing liver fluke disease and liver cancer. Public information on the hazards of raw fish is not enough to deter them from their lifelong habit. In yet another example, people raising pigs or chickens around their homes, a common practice in developing countries, are prone to diseases caught from the animals, zoonotic diseases as they are called, which can then spread to other people.

These examples, illustrating the complex webs of interactions between health and other factors, led to the concept of a broad approach to human health—ecohealth in short—to enable better prevention and mitigation of problems affecting health of individuals and populations. This integrative approach was foreshadowed by, but is generally broader than, the area of environmental health. In the ecohealth approach, health is seen not as a separate entity but as the outcome of interactions between the community, the environment and the economy. This realization leads to the understanding that interventions in an apparently unrelated area can lead to significant outcomes in health, and vice versa. There is therefore a need for collaborative approaches in areas such as agriculture, industry, public service and health. An example is the adoption of an optimal irrigation schedule in rice farming, which has the dual benefit of better yields from water input and a reduction in malaria and other vector-borne diseases [104]. Use of appropriate construction materials, together with better methods of design and construction, could lead not only to cost saving but also to better hygiene. Traditional houses in Central America have walls where triatomid bugs, the vector for Chagas disease, can hide. Intervention by better wall plastering, together with keeping the house tidy, was found to lead to a significant reduction of the infestation [104]. Another example, from outside the area of infectious diseases, is the adoption of better cooking stoves, which both reduces the cost of fuel and leads to fewer respiratory problems arising from smoke.

In the present age of globalization, facile interactions among people from various parts of the world and between people and the environment provide another dimension of complexity to the ecology of infectious diseases. Tourists without immunity to local tropical diseases can be unwittingly infected and carry an exotic disease back into their native countries, with the risk of spreading diseases to new areas. Diseases which can spread directly from people to people, like influenza, carry the highest risk of such a scenario. New, emerging infectious diseases, such as Ebola, SARS and MERS, can potentially spread more rapidly and in various distant places due to the ease of modern travel. In addition to increased efforts in medical science, the adoption of the ecohealth approach can help to mitigate and prevent the risks from such emerging diseases.

6.4   Need for Transformation to Sustainable Development

We have seen that the wilderness is not a passive collection of plants, animals and microbes which we can simply tap for relief from ills, enjoyment of natural beauty or other human purposes. The wilderness cannot be simply consumed without consideration of consequences and sustainability. We have seen examples of nature fighting back against our use of antibiotics and other antimicrobial drugs, triggering a reaction in the form of pathogen resistance. On a broader scale, climate change is nature's response to our way of life, with its intensive use of fossil fuels and production of greenhouse gases. Problems of unsustainability through overexploitation of nature are encountered in deforestation, overfishing of the oceans and poor management of water use. Environmental pollution has been a pressing problem over the past few decades, as pointed out by Rachel Carson in her pioneering work [107]. Carson asserted that our conventional philosophy of seeking control over nature needs to be changed, as it would otherwise lead to our ultimate undoing: 'The "control of nature" is a phrase conceived in arrogance, born of the Neanderthal age of biology and philosophy when it was supposed that nature exists for the convenience of man.'

The problems arising from our use of antibiotics and other drugs against infectious diseases can be seen as consequences of our ingrained attempt to control nature. These problems require a change of mindset, a transformation, so as to be able finally to achieve sustainability. The solutions require the realization that total control may not be possible, or advisable. While elimination of some virulent diseases, such as smallpox in the past, and hopefully poliomyelitis in the future, may be desirable, it may not be a good strategy to aim to eliminate all infectious diseases, either by drugs or by other means. Apart from the fact that this aim may not be economically or technically feasible, it may also draw us into a vicious cycle of encountering resistant and re-emerging or newly emerging pathogens. It may also have long-term implications on the evolution of humans with respect to interaction with other organisms and the environment, including effects on our immunity against diseases. However, this conclusion does not mean that we should simply let nature take its course without our intervention, since this will be an unrealistic call to revert the course of human history. On the contrary, this means that we should learn from past mistakes and take a broader, more sustainable approach to our management of infectious diseases, with judicious use of strategies for prevention and therapy and the ecohealth concept.

In tapping the wilderness for remedies to our illnesses and other useful products, we need to ensure sustainability of the natural sources and the practice in the long run. As already discussed in Chapter 2, Box 2.3, the practice known as biopiracy, the raiding of nature for materials of value without fair compensation for the peoples or nations from whose territories the materials were derived, is now being addressed at the international level. In otherwise regular cases of obtaining materials from nature, we need to consider the sustainability of the practice with regard to conservation of the living species which are the sources as well of the environment in general. When possible and appropriate, production of materials by sustainable farming methods should be preferred to extracting them from nature willy-nilly. When the structures of the natural product compounds are known, and their total or partial synthesis can be achieved more economically compared with extraction from nature, sustainability can be ensured further.

Tapping the molecular wilderness for sustainable development is especially important for tropical developing countries. Sustainable tapping of the molecular wilderness, resulting in remedies for our ailments, yields benefits for people of the whole world. However, tropical developing countries, with the wilderness from which the major global benefits are originated, are also burdened with many infectious diseases and other problems which constitute the dark side of the wilderness. Our sense of humanity and justice therefore demands that we pay special attention to the problems of the tropical developing countries, enabling them to use the tools derived from indigenous biodiversity and more to solve them.

6.5   Tapping Molecular Wilderness Sustainably

Artemis, the Greek goddess of the wilderness, whose Roman equivalent is Diana (Fig. 6.1), is also the goddess of hunting. This is a reflection of the old days, when the wilderness was there for hunting and gathering. We are now in an age in which the wilderness is to be treasured and conserved for sustainable use. Tapping the molecular wilderness in the modern world, therefore, is not hunting for treasures from nature but understanding what it has to offer as known from traditional knowledge and investigating the possibilities of gaining new knowledge and materials to further our health and other purposes. Significantly, we need to do so, not in an exploitative but in a sustainable manner so that the wilderness is not thereby destroyed or compromised but remains much as we found it originally. This requires harder efforts than simple hunting or extracting wanted materials from the wild but eventually will give long-lasting returns, not only for our generation, but also for future generations to come.

Figure 6.1   Artemis is the Greek goddess of the wilderness, hunting, archery, the moon, childbirth and virginity and the protector of young girls, diseases and healing. Her Roman equivalent is Diana, whose sculpture *Diane de Verseilles*, probably the work by Leochares, stands in the Louvre Museum. Public domain licensing by copyright holder.

The new millennium has seen two great international movements concerning the status and future of humankind and nature. The 1992 Earth Summit in Brazil, with the official name of United Nations Conference on Environment and Development (UNCED), attended by 172 governments, gave rise to the declaration of Agenda 21 and other important agreements, including the UN Convention on Biological Diversity [108]. Twenty years later, the Rio+20 or the UN Conference on Sustainable Development renewed the commitment of the Earth Summit and focused on two themes, namely, a green economy in the context of sustainable development and poverty eradication and an institutional framework for sustainable development [109]. This movement, which started mainly with a concern about the environment, has come to the realization that it is inextricably linked with human factors, especially human poverty, and sustainable development of the environment cannot be achieved without human development. Meanwhile, at the turn of the millennium, another UN conference, the Millennium Summit, adopted eight millennium development goals (MDGs) mostly concerned with human development [110]. They include poverty eradication, universal primary education, gender equality, reduction of child mortality, maternal health improvement and combat against infectious diseases, with emphasis on AIDS, TB and malaria. Significantly, the MDGs also include environmental sustainability, with realization of the interdependence with human development. The MDGs were set to be achieved by 2015. However, only a few developing countries have managed to meet these goals satisfactorily, pointing to the necessity for renewed efforts. The post-2015 efforts will now see the congruence of the two movements, resulting in the elaboration of sustainable development goals (SDGs), with integrated actions in economic, social and environmental fields [111]. The post-2015 efforts to reach the SDGs will be collectively made by the various countries and the UN system, hopefully with a better outcome than in the past.

Focusing on living sustainably with the molecular wilderness, we can draw lessons from the past. We should realize that, similar to the SDGs which deal with development on a broader basis, we need to make integrated efforts in various fields. In the medical and health science area, we need better drugs, together with fast and reliable diagnosis of diseases and their potential responses to drugs. We also need better surveillance of diseases and better prevention with the help of vaccines and other means, and adoption of healthy lifestyles which would avoid exposure to pathogens and their vectors and minimize tendencies to develop noncommunicable diseases such as diabetes and cancer. Molecules from nature can help us to achieve these technical goals, but we need to obtain them in a sustainable manner. More importantly, we need to see the broad picture, in which humans and nature interact in a complex manner, and we should make efforts to ensure sustainability in such interactions.

Artemis is not only the goddess of the wilderness and hunting but also of archery, the moon, childbirth and virginity and is the protector of young girls. Significantly for us, she is also the goddess of diseases and healing. In bridging the wilderness with diseases and healing, she covers the message of this book. Together with the joy of nature, the wilderness gives us useful products, many of which we can use for healing diseases. The wilderness also brings many diseases, especially when we trample upon it without due regard. We also need to realize that the wilderness lurks in our modern dwellings and even within us in the forms of pathogens and other causes of diseases. This book has looked at the molecular aspects of the wilderness, which underlie the basis of its risks and benefits. Investigating the molecular aspects of the wilderness, especially at the interface between chemistry, biology and biodiversity, can lead to the discovery of new drugs and the means to combat resistance arising from drug use. Realizing the fragility of biodiversity, we must tread upon the wilderness gently. Tapping the molecular wilderness sustainably should be our virtuous aim.

References

1. *The Oxford Dictionary of Quatations. Seventh Edition.* 2009: OUP.

2. Gillham, N.W., *Organelle Genes and Genomes.* 1994, Wallingford, UK: CABI. 424.

3. Thoreau, H.D., *Walking,* in *The Thoreau Reader. Annotated Works of Henry David Thoreau.* 1862.

4. American Society of Health-System Pharmacists, *Herbal Therapy, Medicinal Plants, and Natural Products: An IPA Compilation.* 1999, Bethesda, MD: American Society of Health-System Pharmacists. vii, 167 p.

5. Bohlin, L., et al., Natural products in modern life science. *Phytochem Rev,* 2010. **9**(2):279–301.

6. Fabricant, D.S., and N.R. Farnsworth, The value of plants used in traditional medicine for drug discovery. *Environ Health Perspect,* 2001. **109**(Suppl 1):69–75.

7. Monod, J., *Chance and Necessity. An Essay on the Natural Philosophy of Modern Biology.* 1971, New York: Alfred A. Knopf.

8. Clardy, J., and C. Walsh, Lessons from natural molecules. *Nature,* 2004. **432**(7019):829–37.

9. Cragg, G.M., and D.J. Newman, Biodiversity: a continuing source of novel drug leads. *Pure Appl Chem,* 2005. **77**:7–24.

10. Dixon, N., et al., Cellular targets of natural products. *Nat Prod Rep,* 2007. **24**(6):1288–310.

11. Imming, P., Molecular targets of natural drug substances: idiosyncrasies and preferences. *Planta Med,* 2010. **76**(16):1794–801.

12. Koh, C.Y., and R.M. Kini, From snake venom toxins to therapeutics: cardiovascular examples. *Toxicon,* 2012. **59**(4):497–506.

13. Warrell, D.A., Snake venoms in science and clinical medicine. 1. Russell's viper: biology, venom and treatment of bites. *Trans R Soc Trop Med Hyg*, 1989. **83**(6):732–40.

14. Hsu, E., Reflections on the "discovery" of the antimalarial qinghao. *Br J Clin Pharmacol*, 2006. **61**(6):666–70.

15. Falkow, S., Molecular Koch's postulates applied to bacterial pathogenicity: a personal recollection 15 years later. *Nat Rev Microbiol*, 2004. **2**(1):67–72.

16. Nobelprize.org. *Emil Fischer: biographical*. 2013. Available from http://www.nobelprize.org/nobel_prizes/chemistry/laureates/1902/fischer-bio.html.

17. World Health Organization, *Traditional Medicine. Fact Sheet No. 134*, W.M. Centre, Editor. 2008: Geneva.

18. Huljic, S., et al., Species-specific toxicity of aristolochic acid (AA) in vitro. *Toxicol In Vitro*, 2008. **22**(5):1213–21.

19. Balachandran, P., et al., Structure activity relationships of aristolochic acid analogues: toxicity in cultured renal epithelial cells. *Kidney Int*, 2005. **67**(5):1797–805.

20. Newman, D.J., and G.M. Cragg, Natural products as sources of new drugs over the 30 years from 1981 to 2010. *J Nat Prod*, 2012. **75**(3):311–35.

21. Farnsworth, N.R., et al., Medicinal plants in therapy. *Bull World Health Organ*, 1985. **63**(6):965–81.

22. Molinski, T.F., et al., Drug development from marine natural products. *Nat Rev Drug Discovery*, 2009. **8**:69–85.

23. Olivera, B.M., and L.J. Cruz, Conotoxins, in retrospect. *Toxicon*, 2001. **39**(1):7–14.

24. Choudhury, B.R., S.J. Haque, and M.K. Poddar, In vivo and in vitro effects of kalmegh (Andrographis paniculata) extract and andrographolide on hepatic microsomal drug metabolizing enzymes. *Planta Med*, 1987. **53**(2):135–40.

25. Lin, F.L., et al., Antioxidant, antioedema and analgesic activities of Andrographis paniculata extracts and their active constituent andrographolide. *Phytother Res*, 2009. **23**(7):958–64.

26. Varma, A., H. Padh, and N. Shrivastava, Andrographolide: a new plant-derived antineoplastic entity on horizon. *Evid Based Complement Alternat Med*, 2011. **2011**:815390.

27. Ligon-Borden, B.L., Biography: Selman A. Waksman, PhD (1888-1973): pioneer in development of antibiotics and Nobel Laureate. *Semin Pediatr Infect Dis*, 2003. **14**(1):60–63.

28. Nobelprize.org. *Sir Alexander Fleming: documentary*. 2013. Available from http://www.nobelprize.org/nobel_prizes/medicine/laureates/1945/fleming-docu.html.

29. Dewick, P.M., *Medicinal Natural Products: A Biosynthetic Approach*. 3rd ed. 2009, Chichester, West Sussex: Wiley. x, 539 p.

30. Clardy, J., M.A. Fischbach, and C.T. Walsh, New antibiotics from bacterial natural products. *Nat Biotechnol*, 2006. **24**(12):1541–50.

31. Handelsman, J., et al., Molecular biological access to the chemistry of unknown soil microbes: a new frontier for natural products. *Chem Biol*, 1998. **5**(10):R245–9.

32. Endo, A., A gift from nature: the birth of the statins. *Nat Med*, 2008. **14**(10):1050–2.

33. Tan, S.Y., and J.K. Zia, Alexandre Yersin (1863-1943): Vietnam's "Fifth Uncle." *Singapore Med J*, 2012. **53**(9):564–5.

34. Bibel, D.J., and T.H. Chen, Diagnosis of plaque: an analysis of the Yersin-Kitasato controversy. *Bacteriol Rev*, 1976. **40**(3):633–51.

35. Cox, F.E., History of the discovery of the malaria parasites and their vectors. *Parasit Vectors*, 2010. **3**(1):5.

36. Garnham, P.C., History of discoveries of malaria parasites and of their life cycles. *Hist Philos Life Sci*, 1988. **10**(1):93–108.

37. World Health Organization. *Trypanosomiasis, human African (sleeping sickness). Fact sheet no. 259*. 2012. Available from http://www.who.int/mediacentre/factsheets/fs259/en/.

38. World Health Organization. *Schistosomiasis*. 2013. Available from http://www.who.int/schistosomiasis/en/.

39. Ehrlich, P., *Ueber den jetzigen Stand der Chemotherapie. Berichte der Deutschen Chemischen Gesellschagt*. 1909. **42**:17–47. Translated in B. Holmstedt and G. Liljestrand, Editors, *Readings in Pharmacology (1963), 286*. 1909.

40. Imming, P., C. Sinning, and A. Meyer, Drugs, their targets and the nature and number of drug targets. *Nat Rev Drug Discovery*, 2006. **5**(10):821–34.

41. Moellering, R.E., and B.F. Cravatt, How chemoproteomics can enable drug discovery and development. *Chem Biol*, 2012. **19**(1):11–22.

42. Cravatt, B.F., A.T. Wright, and J.W. Kozarich, Activity-based protein profiling: from enzyme chemistry to proteomic chemistry. *Annu Rev Biochem*, 2008. **77**:383–414.

43. Vane, J.R. *Nobel Lecture 1982: Adventures and excursions in bioassay; the stepping stone to prostacyclin.* 1982. Available from www.nobelprize. org/nobel_prizes/medicine/laureates/.../vane-lecture.pdf.

44. Rottmann, M., et al., Spiroindolones, a potent compound class for the treatment of malaria. *Science*, 2010. **329**(5996):1175–80.

45. Nok, A.J., Arsenicals (melarsoprol), pentamidine and suramin in the treatment of human African trypanosomiasis. *Parasitol Res*, 2003. **90**(1):71–79.

46. Freire, E., A thermodynamic approach to the affinity optimization of drug candidates. *Chem Biol Drug Des*, 2009. **74**(5):468–72.

47. Polishchuk, P.G., T.I. Madzhidov, and A. Varnek, Estimation of the size of drug-like chemical space based on GDB-17 data. *J Comput Aided Mol Des*, 2013. **27**(8):675–9.

48. Hitchings, G.H., Jr., Nobel lecture in physiology or medicine: 1988. Selective inhibitors of dihydrofolate reductase. *In Vitro Cell Dev Biol*, 1989. **25**(4):303–10.

49. Yuvaniyama, J., et al., Insights into antifolate resistance from malarial DHFR-TS structures. *Nat Struct Biol*, 2003. **10**(5):357–65.

50. Kamali, A., and M. Holodniy, Influenza treatment and prophylaxis with neuraminidase inhibitors: a review. *Infect Drug Resist*, 2013. **6**:187–98.

51. Michiels, B., et al., The value of neuraminidase inhibitors for the prevention and treatment of seasonal influenza: a systematic review of systematic reviews. *PLoS One*, 2013. **8**(4):e60348.

52. Mitchell, J.A., et al., Selectivity of nonsteroidal antiinflammatory drugs as inhibitors of constitutive and inducible cyclooxygenase. *Proc Natl Acad Sci U S A*, 1993. **90**(24):11693–7.

53. Jawabrah Al-Hourani, B., et al., Cyclooxygenase-2 inhibitors: a literature and patent review (2009–2010). *Expert Opin Ther Pat*, 2011. **21**(9):1339–432.

54. Harris, H. *The discovery of penicillin.* 2010. Available from http://www. path.ox.ac.uk/contact/Penicillin.

55. Bellis, M. *The history of penicillin.* 2013. Available from http:// inventors.about.com/od/pstartinventions/a/Penicillin.htm.

56. Elander, R.P., Industrial production of beta-lactam antibiotics. *Appl Microbiol Biotechnol*, 2003. **61**(5–6):385–92.

57. Malik, S., et al., Production of the anticancer drug taxol in Taxus baccata suspensiion cultures: a review. *Process Biochem.*, 2011. **46**:23–34.

58. Paddon, C.J., et al., High-level semi-synthetic production of the potent antimalarial artemisinin. *Nature*, 2013. **496**(7446):528–32.

59. Koch, M.A., et al., Charting biologically relevant chemical space: a structural classification of natural products (SCONP). *Proc Natl Acad Sci U S A*, 2005. **102**(48):17272–7.

60. Rosen, J., et al., Novel chemical space exploration via natural products. *J Med Chem*, 2009. **52**(7):1953–62.

61. Lachance, H., et al., Charting, navigating, and populating natural product chemical space for drug discovery. *J Med Chem*, 2012. **55**(13):5989–6001.

62. Wells, T.N., Natural products as starting points for future anti-malarial therapies: going back to our roots? *Malar J*, 2011. **10**(Suppl 1):S3.

63. Lazo, J.S., and P. Wipf, Combinatorial chemistry and contemporary pharmacology. *J Pharmacol Exp Ther*, 2000. **293**(3):705–9.

64. Furka, A., Combinatorial chemistry: 20 years on. *Drug Discovery Today*, 2002. **7**(1):1–4.

65. Geysen, H.M., et al., Combinatorial compound libraries for drug discovery: an ongoing challenge. *Nat Rev Drug Discovery*, 2003. **2**(3):222–30.

66. Corbett, P.T., et al., Dynamic combinatorial chemistry. *Chem Rev*, 2006. **106**(9):3652–711.

67. Galloway, W.R., A. Isidro-Llobet, and D.R. Spring, Diversity-oriented synthesis as a tool for the discovery of novel biologically active small molecules. *Nat Commun*, 2010. **1**:80.

68. Schreiber, S.L., Target-oriented and diversity-oriented organic synthesis in drug discovery. *Science*, 2000. **287**(5460):1964–9.

69. Dandapani, S., et al., Hits, leads and drugs against malaria through diversity-oriented synthesis. *Future Med Chem*, 2012. **4**(18):2279–94.

70. Lipinski, C.A., et al., Experimental and computational approaches to estimate solubility and permeability in drug discovery and development settings. *Adv Drug Deliv Rev*, 2001. **46**:3–26.

71. Veber, D.F., et al., Molecular properties that influence the oral bioavailability of drug candidates. *J Med Chem*, 2002. **45**(12):2615–23.

72. Erlanson, D.A., Introduction to fragment-based drug discovery. *Top Curr Chem*, 2012. **317**:1–32.

73. Murray, C.W., and D.C. Rees, The rise of fragment-based drug discovery. *Nature Chem*, 2009. **1**(3):187–92.

74. Over, B., et al., Natural-product-derived fragments for fragment-based ligand discovery. *Nat Chem*, 2013. **5**(1):21–8.

75. Wong, F.T., and C. Khosla, Combinatorial biosynthesis of polyketides: a perspective. *Curr Opin Chem Biol*, 2012. **16**(1-2):117–23.

76. Zhang, W., and Y. Tang, Combinatorial biosynthesis of natural products. *J Med Chem*, 2008. **51**(9):2629–33.

77. Rondon, M.R., et al., Cloning the soil metagenome: a strategy for accessing the genetic and functional diversity of uncultured microorganisms. *Appl Environ Microbiol*, 2000. **66**(6):2541–7.

78. Chen, B., et al., Metagenomic profiles of antibiotic resistance genes (ARGs) between human impacted estuary and deep ocean sediments. *Environ Sci Technol*, 2013.

79. Wang, Z., et al., Metagenomic profiling of antibiotic resistance genes and mobile genetic elements in a tannery wastewater treatment plant. *PLoS One*, 2013. **8**(10):e76079.

80. Guantai, E., and K. Chibale, How can natural products serve as a viable source of lead compounds for the development of new/novel anti-malarials? *Malar J*, 2011. **10**(Suppl 1):S2.

81. Colombo, M., and I. Peretto, Chemistry strategies in early drug discovery: an overview of recent trends. *Drug Discovery Today*, 2008. **13**(15-16):677–84.

82. Ball, P., The click concept. *Chem World*, 2007. **April**:46–51.

83. Hou, J., et al., The impact of click chemistry in medicinal chemistry. *Expert Opin Drug Discovery*, 2012. **7**(6):489–501.

84. Lewis, W.G., et al., Click chemistry in situ: acetylcholinesterase as a reaction vessel for the selective assembly of a femtomolar inhibitor from an array of building blocks. *Angew Chem, Int Ed Engl*, 2002. **41**(6):1053–7.

85. Krasinski, A., et al., In situ selection of lead compounds by click chemistry: target-guided optimization of acetylcholinesterase inhibitors. *J Am Chem Soc*, 2005. **127**(18):6686–92.

86. Meek, S.J., et al., Catalytic Z-selective olefin cross-metathesis for natural product synthesis. *Nature*, 2011. **471**(7339):461–6.

87. Schrock, R.R., and A.H. Hoveyda, Molybdenum and tungsten imido alkylidene complexes as efficient olefin-metathesis catalysts. *Angew Chem, Int Ed Engl*, 2003. **42**(38):4592–633.

88. World Health Organization, *Antimicrobial Resistance. Global Report on Surveillance*. 2014, Geneva. 257.

89. World Health Organization. *Antimicrobial resistance. Fact sheet no. 194.* 2013. Available from http://www.who.int/mediacentre/factsheets/fs194/en/.

90. BBC_News, *"Golden age" of antibiotics "set to end."* 2014.

91. Carlson, H.A., et al., Developing a dynamic pharmacophore model for HIV-1 integrase. *J Med Chem*, 2000. **43**(11):2100–14.

92. Schames, J.R., et al., Discovery of a novel binding trench in HIV integrase. *J Med Chem*, 2004. **47**(8):1879–81.

93. Fischbach, M.A., Combination therapies for combating antimicrobial resistance. *Curr Opin Microbiol*, 2011. **14**(5):519–23.

94. Worthington, R.J., and C. Melander, Combination approaches to combat multidrug-resistant bacteria. *Trends Biotechnol*, 2013. **31**(3):177–84.

95. Aminov, R.I., The role of antibiotics and antibiotic resistance in nature. *Environ Microbiol*, 2009. **11**(12):2970–88.

96. Chait, R., K. Vetsigian, and R. Kishony, What counters antibiotic resistance in nature? *Nat Chem Biol*, 2012. **8**(1):2–5.

97. Cowman, A.F., et al., Amino acid changes linked to pyrimethamine resistance in the dihydrofolate reductase-thymidylate synthase gene of Plasmodium falciparum. *Proc Natl Acad Sci U S A*, 1988. **85**(23):9109–13.

98. Peterson, D.S., D. Walliker, and T.E. Wellems, Evidence that a point mutation in dihydrofolate reductase-thymidylate synthase confers resistance to pyrimethamine in falciparum malaria. *Proc Natl Acad Sci U S A*, 1988. **85**(23):9114–8.

99. Rieckmann, K.H., The in vitro activity of experimental antimalarial compounds against strains of P. falciparum with varying degrees of sensitivity to pyrimethamine and chloroquine. *WHO Tech Rep Ser*, 1973. **529**:58.

100. Rieckmann, K.H., A.E. Yeo, and M.D. Edstein, *Activity of PS-15 and its metabolite, WR99210, against Plasmodium falciparum in an in vivo-in vitro model.* Trans R Soc Trop Med Hyg, 1996. **90**(5):568–71.

101. Yuthavong, Y., et al., Malarial dihydrofolate reductase as a paradigm for drug development against a resistance-compromised target. *Proc Natl Acad Sci U S A*, 2012. **109**(42):16823–8.

102. NIAID. *Emerging and re-emerging infectious diseases.* 2012. Available from http://www.niaid.nih.gov/topics/emerging/pages/list.aspx.

103. Intergovernmental_Panel_on_Climate_Change, *Climate Change 2014. Fifth Assessment Report.* 2014: IPCC.

104. Charron, D.F., *Ecohealth Research in Preactice. Innovative Applications of an Ecosystem Approach to Health*. 2012: International Development Research Center.

105. Forget, G., and J. Lebel, An ecosystem approach to human health. *Int J Occupat Environ Health*, 2001. **7**(2) (Supp):S3–S38.

106. Lebel, J., *An Ecosystem Approach*. 2003, Ottawa: International Development Research Center.

107. Carson, R., L. Darling, and L. Darling, *Silent Spring*. 1962, Boston, Cambridge, MA: Houghton Mifflin; Riverside Press. 368 p.

108. UN. *Earth Summit. UN Conference on Environment and Development (1992)*. 1997. Available from http://www.un.org/geninfo/bp/enviro.html.

109. UN. *Future we want. Outcome document*. 2012. Available from http://sustainabledevelopment.un.org/futurewewant.html.

110. UN. *We can end poverty. Millennium development goals and beyond 2015*. 2014. Available from http://www.un.org/millenniumgoals/.

111. UN. *Sustrainable development goals*. 2014. Available from http://sustainabledevelopment.un.org/?menu=1300.

Glossary

Acetylcholine	An organic molecule serving as transmitter of nerve signals in humans and many species.
Alkaloids	Nitrogen-containing, mostly basic, compounds derived mostly from amino acids. Many have strong biological (drug, narcotic or toxic) activities.
Amino acid	An organic acid containing an amino group, an important component of living organisms both in itself and a building block of proteins.
Aminoglycosides	A group of compounds, many of which are antibiotics including streptomycin and gentamycin.
Andrographolide	Agent derived from the herbaceous plant *Androgaphis paniculata*, with many pharmacological properties.
Antibiotic	Agent which kills or slows the growth of microorganisms, usually made by a microorganism, but also by other living species.
Antifolates	A group of drugs which work by inhibiting folate metabolism.
Artemisinin	Antimalarial originally discovered in China, found in sweet wormwood.

Aspirin	Agent relieving fever aches and pains, found in willow bark but now made synthetically.
Bacteria	A large group of microbes grouped as prokaryotes which are single cells without a nucleus.
Beta-lactams	A group of compounds, many of which are antibiotics, including penicillin.
Bioavailability	The extent to which a drug is available to the target molecule or tissue in the body after administration.
Biodiversity	Degree of variation of life, the richness of which indicates the health of an ecosystem.
Biology	Study of life forms and processes, including interaction among themselves and with the environment.
Biomolecules	Molecules made by living organisms, including proteins, carbohydrates, lipids and nucleic acids, serving particular functions in life processes.
Biopiracy	Illegitimate appropriation of materials and intellectual property of local communities.
Bioprospecting	Survey and use of materials from biodiverse environment.
Biosynthesis	Making of a molecule by living organisms
Caffeine	Stimulatory agent found in coffee, tea and other plants.
Chagas disease	Tropical parasitic disease found in Central and South America caused by a protozoon, *Trypanosoma cruzi*, and spread by an insect known as the kissing bug. Also called American trypanosomiasis.
Chemical genetics	Use of chemicals, mostly small molecules, to study the functions of cells at the gene level.

Chemical library	Collection of chemical compounds used in screening for drugs or in chemical synthesis.
Chemical space	Ensemble of all possible molecules, each of which is located in the space according to its molecular properties.
Chemistry	Study of substances, both from nature and man-made, including their identification, properties and interactions with one another
Chloroplast	Membranous structure in cells of plants and some microorganisms responsible for photosynthesis, or making of sugars using sunlight as energy source.
Chloroquine	An antimalarial which had been very effective until the sixties, when widespread resistance occurred.
Chromatography	A technique for separation of mixtures.
Climate change	Change of climate which can be the result of human activities
Combinatorial chemistry	Making of a large number of molecules by combination of various components.
Coumarins	A group of natural products from plants with a wide range of biological functions.
Culture	Cultivation of microorganisms or cells from tissues or organisms in an artificial medium containing nutrients and a suitable environment.
Cyclooxygenases	A group of enzymes involved in the production of prostaglandins and thromboxanes, responsible for the inflammation and clotting processes respectively.
Dalton	Unit of atomic and molecular weight, based on the ratio of the real weight and the weight of the hydrogen atom.

Digoxin	An agent from foxglove plant used to treat various heart conditions.
Dihydrofolate reductase	An enzyme involved in production of many metabolites, and is a target for a group of drugs called antifolates.
DNA	Deoxyribonucleic acid, the genetic component of various organisms.
Drug	Substance used to cure or alleviate symptoms of diseases.
Drug resistance	Reduction of effectiveness of a drug in treating a disease.
Drug target	Target of action of a drug, usually a molecule or cellular structure where the drug exerts its primary effect.
Drug tolerance	Ability to tolerate drug effect.
Druggability	Propensity of a biological target to be effectively altered functionally by a drug.
Ecology	A branch of biology concerning interactions between living species with one another and with their environment.
Ecosystem	Community of plants, animals and microorganisms which interact with one another and with the environment.
Emerging disease	Disease which threatens humans from recent outbreaks.
Enzyme	Biomolecule catalyzing biological and other reactions in life processes.
Ethnobotany	Study of plants used in traditional medicine.
Ethnomedicine	Study of practice of ethnic groups for healing sickness and maintaining health, usually handed down by traditional beliefs over generations.
Ethnopharmacy	Study of the use of traditional drugs by ethnic groups

Evolution	Change in genetic characters of living organisms over successive generations so as to gain fitness in surviving and reproducing in the environment.
Extremophile	An organism which thrives in extreme conditions, such as hot springs and ocean depths.
Fansidar	A combination of sulfadoxine and pyrimethamine, still remaining mostly effective in Africa, although there is widespread resistance in other areas of the world.
Fermentation	Process for growing microorganisms for production of compounds or cells.
Flavonoids	A class of natural products widely distributed in plants, fulfilling many functions including being pigments.
Foxglove	A perennial herbal plant in the genus Digitalis, containing agents with medical and poisonous properties.
Fragment-based drug discovery	Discovery of drugs from the study of the components of their molecules.
Gene knockout	Disabling a gene by biochemical or genetic techniques.
Genome	All genes of an organism, including those with known and unknown functions.
Genomics	Study of all genes of an organism and their functions.
Global warming	Warming of the whole world, which can be caused by human activities.
Greenhouse gas	Gas that traps heat in the atmosphere.
Herbal drug	Drug made from herbs, usually from instruction derived from traditional medicine.

High-throughput screening	Drug discovery process based on rapid examination of drug effects of a large number of compounds by an automatic process.
Hit compound	A compound giving positive results in initial drug screening.
Immunomodulating properties	Properties of some agents to modify the immune response, including suppressing and potentiating the response.
Infectious disease	Disease caused by a pathogen, which can be a virus, bacterium or other organisms.
Influenza	Respiratory disease caused by the influenza virus.
Intellectual property rights	The rights of inventors over their inventions, which can be licensed out for use by others with mutual agreement.
Isoprene	A molecule consisting of five carbons, forming the basic unit for terpenoids, steroids and other natural products.
Lead compound	A compound with properties which might be therapeutically useful, usually selected from many hit compounds.
Ligand	A substance, usually a small molecule, which binds with a biomolecule, causing such changes as activation of its function.
Liposomes	Tiny bubbles made of lipids which can contain drugs to be delivered to cells and tissues.
Malaria	Disease with fever and other symptoms cause by protozoa of the genus Plasmodium.
Mass spectrometry	A technique for determining molecular mass and details by finding mass to charge ratios.
Metabolism	Series of reactions in an organism leading to various products, or metabolites.

Metabolite	A product of metabolism, which can also be the starting point for other metabolites.
Metagenomics	Study of genetic material recovered directly from the environment.
Micelles	Tiny aggregates of soap-like molecules.
Microbiome	All microorganisms in a location, such as the gut or the skin of an animal.
Microorganism	Very small, microscopic organism, which may be single cell or multicellular.
Mitochondrion (*plural*: mitochondria)	Membranous structure within eukaryotic cells, including plants, animals and fungi) responsible for producing energy.
Molecule	The basic unit of a chemical substance, consisting of two or more atoms held together by chemical bonds.
Mutation	Change in a gene due to change in its sequence.
Nanoparticles	Small particles with dimension of a few nanometers (10^{-9} m)
Natural product	Chemical compound produced by a living organism found in nature, often with pharmacological or biological activity.
Nicotine	An addictive natural product from the nightshade family of plants and also in tobacco.
Opiates	Narcotic alkaloid drugs, including morphine, found as natural products in the opium plants. Other drugs with similar properties, binding with the sane receptors, are called opioids.
Oseltamivir	An anti-influenza drug, also known commercially as Tamiflu.
Papain	Protein-digesting enzyme found in the skin of raw papaya fruit.
Parasite	An organism that lives in or on a host and derives benefit from it at its expense.

Pathogen	An infectious agent which can produce disease.
Penicillin	An antibiotic originally derived from Penicillium fungi, classified as a beta-lactam.
Peptide	A biomolecule composed of a small number of amino acids, say up to 50, joined together as a chain.
Pharmacodynamics	Study of biochemical and physiological effects of a drug on a human or any organism taken by or exposed to it.
Pharmacognosy	Study of medicines derived from natural sources, including their physical, chemical, biochemical and biological properties.
Pharmacokinetics	Study of the fate of a drug in a human or any organism taken by or exposed to it.
Pharmacology	Study of interactions between an organism and drugs or other chemical substances
Pharmacopoeia	A book containing directions for the identification of medicines, usually published by a government or a medical or pharmaceutical society.
Pharmacophore	A part of the molecule of a drug responsible for the drug's action, often through binding with the receptor triggering biological response.
Phenylpropanoid	A diverse group of natural products synthesized by plants from phenylalanine.
Pheromone	Chemicals given off by insects and various other organisms in tiny amounts to transmit alarm, indicate food sources or attract members of the opposite sex.
Photosynthesis	Synthesis of sugars by plants and other organisms from carbon dioxide and water, using the energy from sunlight which is stored as chemical energy in the sugar molecules.

Polyketides	A large class of natural products from microorganisms, plants and animals, basically formed from coupling of 2-C (acetyl) units.
Probiotic	A microorganism which is believed to provide health benefit when consumed.
Protein	A biomolecule consisting of around 100 or more amino acids joined together as a chain, or as a group of chains.
Protozoa	A diverse group of unicellular eukaryotic organisms, which include both free-living and parasitic forms.
Quinine	Bitter principle found in the bark of cinchona tree, with antimalarial, anti-inflammatory and pain-killing properties.
Receptor	A protein molecule or more complex structure on the surface or inside a cell, which receives signals from outside through binding with a drug or other molecules, resulting in further biochemical or physiological changes.
Rhizobium	A group of bacteria living symbiotically with plants in their roots and turn nitrogen into nutrients for the plants.
RNA	Ribonucleic acid, derived in life process from DNA, with various functions including synthesis of proteins.
Schistosomiasis	A disease caused by a parasitic worm of the genus Schistosoma, living in the blood and internal organs of humans and various animals.
Sleeping sickness	A parasitic disease caused by protozoa of the genus Trypanosoma and transmitted through the bites of tsetse flies, characterized by fever, headaches and joint pains, followed by confusion, numbness and trouble sleeping. Also called African trypanosomiasis.

Statins	Drugs used for lowering of cholesterol levels, originally derived from microbial fungi.
Steroids	A large class of organic compounds with various biological functions consisting of isoprene units joined together to make four fused rings and side chains.
Streptomycin	An antibiotic derived from the bacterium Streptomyces, classified as an aminoglycoside.
Substrate	A compound undergoing a reaction which is catalyzed by an enzyme.
Sustainability	Ability to be used without destruction, depletion or permanent damage.
Symbiosis	Close and often long-term interaction between two or more different species, resulting in mutual benefits.
Taxol (Paclitaxel)	A natural product from plant which is a mitotic inhibitor used in cancer treatment.
Terpenoids	A large class of natural compounds consisting of isoprene units (5-carbon molecules) joined together in various fashions.
Toxin	A poisonous substance produced by a cell or an organism to debilitate or kill another cell or organism.
Trypanosomiasis	A group of diseases caused by protozoa of the genus Trypanosoma.
Vaccine	A substance which protects a person or an animal against a disease.
Venom	Toxic substance used by snakes, insects or other animals against their enemies or preys.
Virus	A small infectious agent which replicates inside the cell it infects.

| X-ray diffraction | A technique for determination of molecular structure of crystalline substances, such as some proteins. |
| Zanamivir | An anti-influenza drug, also known commercially as Relenza. |

Index

plasmids 86
Plasmodium falciparum 6, 84, 90
pneumococcal pneumonia 48
poison mushroom 3
poliomyelitis 48, 104
polyene 36
polyketides 36, 79
polyketide synthase 79
predator 3
prey 3
primary metabolism 35
prophylactics 49
prostaglandins 57, 63
protein profiling 57
proteins 37
proteomics 56, 57
public health system 87
pufferfish 3
pyrazinamide 92
pyrimethamine 54, 84, 90, 92, 94
pyrimethamine-sulfadoxine 6, 83

quinine 6
quinine 37
quinolones 70

raltegravir 92
Ramayana 50
random screening 65
receptors 52
Relenza (zanamivir) 63
resistance surveillance 87
reverse pharmacology 56
reverse transcriptase 92
rhinoceros horns 8
Rhizobium 4
ribosomes 52
rifampicin 92
rifamycin 79
Rio+20 107
ritonavir 92
RNA 52
Ross, Ronald 45

salamanders 10
salvarsan 51
SARS 48, 98
schistosomes 4, 47
schistosomiasis 47
Schopenauer, Arthur 43
Schrock, Richard R. 77
SDGs 107
sea squirts 29
secondary metabolism 36
secondary metabolite 79
semisynthetic drugs 59
Senna 6, 7
severe acute respiratory syndrome
    (SARS) 95, 98
Sharpless, Barry 75
skunks 9
sleeping sickness 46, 47, 60
smallpox 48, 98, 104
snake venoms 2, 9
solid-phase peptide synthesis 73
South American frogs 3
spiroindolones 57, 75
sponges 28
spongothymidine 29
spongouridine 29
statins 27
steroids 37
stevioside 26
Stewart, William 98
Streptococcus faecium 54
Streptomyces 69, 70, 95
streptomycin 30
Strophantus gratus 26
structure–activity relationships 62
substrate 12, 52
suicide substrate 60
sulfadoxine 94
sustainable development goals
    107
sweet wormwood 5, 6, 7
symbiosis 4, 10
symptoms 46

#0001 - 220617 - C26 - 229/152/9 [11] - CB - 9789814613590